THE CAMBRIDGE
PHILOSOPHICAL SOCIETY

A History, 1819-1969

A. Rupert Hall

Cambridge Philosophical Society 1969

Contents

Acknowledgements

Acknowledgement is made for permission to reproduce portraits and photographs as follows :—

To the Cambridge News:
Sir John Cockcroft

To the Master and Fellows of Christ's College:
E. W. Hobson

To Mr. G. P. Darwin:
Sir George Darwin

To Mrs. N. Fox:
H. Munro Fox

To the Master and Fellows of Jesus College:
E. D. Clarke

To the Master and Fellows of St. John's College:
J. C. Adams, H. F. Baker, J. S. Henslow, G. D. Liveing

To the Master and Fellows of Magdalene College:
William Farish

To Mrs. L. Meitner-Graf:
Niels Bohr

To the Master and Fellows of Peterhouse:
William Hopkins

To the Master and Fellows of Sidney Sussex College:
C. T. R. Wilson

To the Master and Fellows of Trinity College:
G. B. Airy, F. W. Aston, C. Babbage, F. M. Balfour, J. Challis, J. Cumming, Sir R. Glazebrook, J. Clerk Maxwell, S. Ramanujan, William Whewell

To the University and its Departments:
C. C. Babington, Sir J. Barcroft, William Bateson, J. Willis Clark, Sir M. Foster, Sir F. Gowland Hopkins, H. F. Newall, A. Newton, Lord Rutherford, Sir A. C. Seward, Sir A. Shipley.

Prefatory Note

For encouragement and help in preparing this brief history I am much indebted to Sir Frank Engledow, Sir William Hodge (who greatly improved its final pages), Mr. J. A. Ratcliffe (its 'onlie begetter') and Professor F. J. W. Roughton. I am also grateful to a number of College librarians who kindly furnished information about their old members. Without materials provided by the Philosophical Society's Librarian, Miss J. E. Larter, including all the illustrations, it could not have been written at all. I thank all of these, and absolve them of all responsibility for what I have done with their contributions.

The last history of the Philosophical Society was prepared by John Willis Clark as his valedictory Address on retiring from its Presidency (on 27 October 1890) and published in Volume 7 of the *Proceedings*. As this account dealt very fully with the origins of the Society and the first few decades of its existence, I have made this part of my story the more brief.

There are necessarily many other omissions from this narrative. Some are deliberate. I have hardly touched on the outstanding achievements of the most distinguished Fellows where, so far as I know, these were unconnected with the Philosophical Society. I have made only passing allusions to the scientific work of living persons, though some of these have been Fellows for half a century, and I have accordingly attempted no systematic account of the scientific work published by the Society and its Fellows during the years since the last war. Moreover, some work of the distant past and of the inter-war years has, for lack of space, not received the attention its importance really merits. The Biographical Notes at the end may, I hope, repair some omissions from the narrative.

For other sins both of omission and commission I can only plead that I am not William Whewell, and (with my sincere regrets) proffer Dr. Johnson's apology: 'Ignorance, Madam, pure ignorance.'

The names *Transactions* and *Proceedings* (unless qualified) always refer to the Philosophical Society's publications.

A. R. HALL

*Professor of the History of Science
and Technology
Imperial College, London*

CHAPTER ONE

The Foundation of the Philosophical Society

THE FOUNDATION of the Cambridge Philosophical Society in 1819 was the first positive step taken in modern times towards the emergence of Cambridge University as a great centre for teaching and research in science. It was the first move from almost a century of indolence and dullness during which, if Newton's name had been revered, his own example of relentless intellectual activity had rarely been followed. Admittedly, in the second decade of the nineteenth century the Cambridge scientific scene revealed little of brilliance or promise. The best that can be said of the University is that its condition had been even more deplorable a generation or so before, and that the consumption of port in Combination Rooms has been exaggerated by legend. The Statutes which both University and Colleges were bound to observe, including one requiring the Regius Professor of Physic to lecture only upon the writings of Hippocrates and Galen, were hopelessly irrelevant to the needs of the day. In this fact the idleness, corruption and shameless self-seeking characterising every aspect of English life during the latter part of the eighteenth century had discovered a ready excuse for the breach of every Statute not protected by self-interest or long custom, and the neglect of every natural obligation. Few among teachers and taught made any pretence of doing the work expected by earlier and later centuries of the various members of a University. Professors hardly ever lectured. They, and even Heads of Colleges, were commonly absentees and pluralists. A College Fellowship was normally regarded as conferring an agreeable membership of a club until such time as the Fellow passed on to a church living and marriage. As for the undergraduates, who were dismally few in number, their real interest in shooting, riding, boating, violence, drinking and gambling was hardly at all obstructed by

1

the ultimate necessity to scramble in parrot Latin through some meaningless exercises of medieval descent in order to qualify for a pass degree. Yet there were some undergraduates who studied hard, just as there were a few Tutors to encourage and guide them. As a Scottish visitor to Cambridge wrote in 1805, to be a high wrangler 'requires reading that in Scotland we have hardly any notion of. If there are greater instances of idleness in English seminaries, there are likewise more astonishing proofs of application'. Certain parts of mathematics were, indeed, virtually the only subjects seriously read in the Colleges, and properly examined by the University. Yet then as now the hard readers were not always the men of greatest promise; Charles Darwin recollected how he had as an undergraduate (1827–30) indulged his passions for hunting, shooting, and collecting insects; 'and been *quite idle.*'

If the professors had done their duty, there would have been plenty of scientific activity in Cambridge. Besides the Regius Chair in Physic (held by the silent Isaac Pennington) there was the Lucasian Chair of Mathematics from which Newton had once lectured, now also quite silent for over half a century, and the Lowndean professorship which had apparently never been used for teaching. During the long tenure of Anthony Shepherd the Plumian Chair of Astronomy had been useless, but a few lectures had been given since 1796. As titular Professor of Botany Thomas Martyn had attempted to teach, since he had the right to charge fees for so doing, until he retired to Bedfordshire in 1798, retaining the professorship for a further twenty-seven years. The Professor of Anatomy, Busick Harwood, was compelled to lecture by the terms of a parliamentary grant he received. Although some of the Woodwardian Professors of Geology attended to Woodward's specimens—not always advisedly—none ventured to lecture until Adam Sedgwick came to the Chair in 1818. On the other hand Richard Watson, titular Professor of Chemistry, had given lectures as did Isaac Milner and Francis Wollaston, the first two occupants of the Jacksonian Chair of Natural Philosophy founded in 1782. Because Wollaston gave a course on chemistry illustrated by many experiments William Farish, as Professor of Chemistry from 1794 to 1813, devoted his lectures to the applications of that science in technology and industry.

The difficulty of securing an audience alleged by the Professors was not wholly imaginary since so many branches of knowledge to which Chairs were attached had no place in University studies; conversely, the Professors of mathematical subjects could gracefully excuse their inattention to that knowledge which the University did examine by pointing to the excellence of the College system of tuition. Most students probably went no further than a little Euclid and the extraction of roots; the best were searchingly tested on fluxions and the early sections of Newton's *Principia*. Unfortunately the topics to which the genuine mathematical abilities of the high wranglers were thus applied bore about as little relation to the science of Laplace and Gauss as did the ancient Chinese classics.

That Cambridge University needed reform had long been perceived. Reform had even been attempted, and in the years after 1800 conscience began to induce the new appointees to Chairs, in turn, to begin to lecture. It is obvious that when the state of science in the English Universities was so low—for Oxford was no better than Cambridge—English science could hardly be in a thriving condition. Much might be expected from the abilities of the dissenters who, like Joseph Priestley, were excluded from the Universities by the religious tests; from aristocrats like Henry Cavendish (a Fellow-Commoner at Cambridge), or from the active Midland manufacturers who formed the Lunar Society. But only un-thinking and selfish men could have imagined, about 1800, that all such efforts compensated for the Universities' negligence. And the general ills of English society affecting them affected the Royal Society in its leadership of British science hardly less. The long presidency of Sir Joseph Banks aroused anxiety in liberal, innovating minds. Despite opposition, dissident specialist societies were formed. A generation of criticism and malaise was to culminate about 1830, in a conviction widespread among the younger devotees of science that its state in England was far from healthy.

The complaints of the reformers of 1830 were perhaps a trifle hysterical, although scientists of other nations also felt a sense of unjust neglect. In England the reform movement had already begun, initiated indeed by the foundation of the Royal Institution as early as 1798; the nation of Dalton, Davy, Young, Wollaston, Lyell, Bell, Brown and young Faraday was not destitute of scientific accomplishment. In 1830 an era of great distinction for British science was commencing.

In this era and in the movement of reform that preceded it the Cambridge Philosophical Society played a large part.

Adam Sedgwick 1785–1873

The first scientific group at Cambridge was to have included Isaac Newton (Trinity) and Henry More, the philosopher (Christ's), but came to nothing because of 'the want of persons willing to try experiments, he whom we chiefly relied on, refusing to concern himself in that kind' as Newton wrote (1685). Much later, in 1784, a *Society for the Promotion of Philosophy and General Literature* was established, which lasted only a couple of years though its membership included many notable representatives of science in Cambridge, among them Isaac Milner, William Farish, Samuel Vince the Plumian Professor, Thomas Martyn, Smithson Tennant, and F. J. H. Wollaston.

The third attempt succeeded. Two men were the effective creators of the Cambridge Philosophical Society: the elder, Adam Sedgwick (1785–1873) a massive, craggy, and often temperamental Yorkshireman, had been elected on promise to the Woodwardian Chair of Geology only in the previous year, when he was much less familiar with the science than was the rival candidate. He had been fifth Wrangler in 1805 and a Fellow of Trinity since 1810, having prepared so hard for the competition that his health was (it is said) permanently impaired. As a professor he applied the same industry and conscientiousness. He soon equipped himself to give his first course of lectures, and published in the first part of the Society's *Transactions* (1821) his own first paper 'On the Geology of Cornwall and Devon'. During a long active life Sedgwick was a pillar of the Geological Society as well as of Cambridge science, and a foremost interpreter of the stratigraphy of British rocks. As might be expected his position became, in later life, somewhat conservative, for he was already too old to admit completely the uniformitarianism of Lyell (*c.* 1830), and the theories of Darwin he always detested.

Charles Robert Darwin had been Sedgwick's pupil on an excursion to North Wales in the summer of 1832, when he learnt several lessons about science never forgotten, but he owed far more to the other founder of the Cambridge Philosophical Society, John Stevens Henslow (1796–1861), whose friendship (Darwin wrote), 'influenced my whole career more than any other' and whose character he valued in the highest terms. It was Henslow who sent Darwin on the *Beagle*, and who in 1835 presented extracts out of Darwin's letters from South America to the Society.* In 1819, however, Henslow had only just taken a rather undistinguished degree, having devoted much of his time at St. John's to insects and shells, as well as to the lectures of James Cumming on chemistry and those of Edward Daniel Clarke on mineralogy. These real enthusiasms did not of course count in the Senate House Examination. At first, in close association with Sedgwick, Henslow went deeply into geology (his first paper in the Society's *Transactions*, 1822, was on the rocks of Anglesey) so that he was universally regarded as being highly qualified (by the standards of the age) to succeed E.D. Clarke in 1822. Nevertheless a great scandal was provoked. As the titular Professorship of Mineralogy was governed by no Statutes, the Heads of Houses claimed the right to compel the Senate to elect one or other of their two nominees. Henslow

---

* They were privately printed in the same year as a small pamphlet, which was reissued by the Philosophical Society in 1960.

J. S. Henslow 1796–1861

was one of these. Despite his fitness, the Senate on principle voted for a third man, whereupon the Vice-Chancellor declared Henslow appointed. Before this dispute, in which Adam Sedgwick was one of the hottest opponents of the Heads of Houses, had been decided legally in their favour, Henslow in 1825 had been nominated by the Crown King's Professor of Botany, temporarily retaining the titular and unpaid Chair of Mineralogy but allowing its duties to lapse. Henslow confessed he was not so well qualified for the salaried Chair, but thought he knew as much of botany as any one else in the University. Not many years later he moved to a living in Suffolk, though continuing to give annual courses of lectures. Darwin, who particularly emphasised Henslow's intense piety, nevertheless maintained their friendship through the emergence of the *Origin of Species*.

Although chance obviously intervenes in the succession to Chairs, it is hardly an accident that the founders of the Cambridge Philosophical Society were not mathematicians, since the condition of the University inspired some activity in chemistry, natural history and geology, while the state of Cambridge mathematics inspired professors of that subject to utter passivity. George Peacock was the first mathematical professor to lecture, as late as 1837. The Preface to the first volume of the Society's *Transactions* was indeed more tactful than accurate in stating that the 'various departments of Mathematical and Philosophical Learning have long occupied a distinguished place in the system of Education adopted in the University of Cambridge . . . Hence, as might naturally be expected, men, eminent for their proficiency in those branches of learning, have never been wanting in this University.' Certainly there was reason to believe that there were enought devotees of mathematics and science in Cambridge to render a Society viable, but the very arguments used on its behalf in this Preface—the absence of facilities for discussion and publication, the need to encourage young men to stick to their work after taking their degrees—indicate that though matters were improving, they were still very imperfect.

The idea of a Philosophical Society was discussed by Sedgwick and Henslow in the course of a geological tour of the Isle of Wight made in the Easter vacation of 1819. According to Henslow's biographer their first thought was for a Corresponding Society; evidently it would not be limited to mathematical studies, and indeed the purpose of the Society was restated, soon after its foundation, to read: 'That this Society be instituted for the purpose of promoting scientific enquiries and of facilitating the communication of facts connected with the advancement of Philosophy *and Natural History*' (my italics). The same breadth of interest was emphasized again in the Preface to the first volume of *Transactions*.

However, this was some little while ahead. Sedgwick and Henslow took no immediate steps, save to sound out their friends, notably E. D. Clarke (1769–1822), titular Professor of Mineralogy, whose reputation was based on his extraordinary travels, and enormous collections brought back (sometimes, virtually looted) from exotic places. At last a public meeting was called at the University Library for 2nd November 1819. Besides those already mentioned the invitation was signed by

7

E. D. Clarke 1769–1822

several Heads of Houses, the Registrary and the Librarian, by the Regius Professors of Medicine and Greek, by the Lucasian Professor (Robert Woodhouse, 1773–1827, the first teacher of calculus in England), and by the professors of Arabic and Chemistry, by George Peacock and William Whewell. There were 33 signatories in all. The inauguration of the Society has been described at some length by John Willis Clark; it is enough here to note that it was unanimously resolved on 2nd November to create the Society, and that a committee appointed to draw up Regulations was so expeditious that on the 15th the Regulations were adopted. The first ordinary meeting of the Philosophical Society was held on 13th December 1819 in the Museum of the Botanic Garden, the chief event being E. D. Clarke's speech on the objects of the Society, in which he spoke of the need for a meeting-place and Museum, and asked the University to support the Society's *Transactions*, as it was indeed to do. 'There is every reason to hope', he concluded, 'that the Graduates of this University, who associated for the Institution of the Cambridge Philosophical Society, by their assiduity and diligence in its support, and by their conspicuous zeal for the honour and well-being of the University, will prove to other times that their Lives and their Studies have not been in vain.'

By the end of 1820 there were 171 Fellows, and £300 had been invested. Of the senior members of the University, wrote Sedgwick, some laugh, some think the proceedings subversive of good discipline, a much larger number look on with indifference, and a few 'are among our warm friends.'

If a proposal made by E. D. Clarke in this same year had been accepted the Society would have been known as the 'Cambridge Philosophical *and Literary* Society'. Though its interests were always broadly interpreted through the Society's first half century, the great majority of papers was devoted to mathematical, scientific or technical topics. Only rarely does one range outside. In 1823, for instance, a letter from George Peacock was read about a recently discovered papyrus containing lines from the *Iliad*, and other contributions on archeological matters came in from time to time. Quite a number of papers are concerned with the history of mathematics and science, De Morgan's interest here being maintained by Maxwell and Larmor. Whewell took up the history of architecture, a subject later developed before the Society by Robert Willis and John Willis Clark. Whewell also devoted a lecture to the economic theory of Ricardo, and others to philosophy. About 1860 a Fellow named J. W. Donaldson published a group of papers in the *Proceedings* dealing with points of classical scholarship, and these serve as a reminder that not all even of the officers of the Society were most distinguished for excellence in science, though no doubt all had done creditably in the Mathematical Tripos. But the concentration of interest was always on pure and applied science and later suggestions having the same purport as Dr. Clarke's proposal never succeeded in altering the scientific character that the Philosophical Society had always, from the first, been intended to possess.

CHAPTER TWO

Some early Fellows
of the Society

URING THE first decades of its existence a number of
men of diverse scientific attainments, some of them distinguished on the national scene as well as in Cambridge, concerned themselves actively with the affairs
of the Philosophical Society. Its first President, William Farish
(1759–1837), who had been promoted to the Jacksonian Chair
from the professorship of chemistry in 1813, was one of the best
scientists and teachers in the University at this time. In the
first learned paper delivered to the Society (20 February 1820)
he described the isometric projection for engineering drawing
he had devised in order to record permanently the various
mechanical models employed in his course of lectures on machines, these models being assembled as required from a set of
interchangeable parts rather like the modern Meccano. Some of
these axles, gears, eccentrics and so on are still preserved in the
Department of Engineering. In subsequent years Farish presented a few other papers on engineering topics. The most
distinguished of the other biennial Presidents during the first ten
years—for Sedgwick himself was in office from 1831–33—was
James Cumming (1777–1861), Professor of Chemistry since
1815, President 1825–27. In the later period of his life Cumming
was wholly devoted to electrical science. Following the discoveries of Oersted in 1820, he gave the first account in Cambridge, with experiments, of 'The effects of the galvanic fluid on
the magnetic needle' at a meeting of the Society on 2 April
1821. On several subsequent occasions Cumming addressed the
Society again on electromagnetism and electrodynamics, as
well as quite unrelated topics, but his main research interest
was thermoelectricity. At these early meetings of the Philosophical Society the topics of papers were highly varied, ranging
from internal calculi (another of Cumming's interests) to the

William Farish 1759–1837   James Cumming 1777–1861

mathematical principles of chemical philosophy, from an ancient sarcophagus to the hydrogen gas-engine described by William Cecil of Magdalene, with geological and palaeontological subjects perhaps the commonest of any one sort.

In 1820 papers appear given by younger men of great subsequent fame. Charles Babbage (1792–1871), author of 'The Calculus of Functions' (May 1820) was one of the original Fellows but never an officer of the Philosophical Society, his connections with the Royal Astronomical Society (founded in 1820) and the Statistical Society (1834) being far more intimate. Several further papers were read by or for him. His disrespect for tradition was manifest in his forming (with Herschel and Peacock) the Analytical Society whose aim was to oust fluxions; and again in 1830 in writing *Reflections on the Decline of Science in England*, an outspoken attack upon the Royal Society. Yet Babbage was traditionalist enough to give no lectures as Lucasian Professor (1828–39). His conception of a calculating machine, or rather machines, is as well known as the disputes to which it gave rise.

John Frederick William Herschel, so close an associate and contemporary of Babbage's that they were born and died in the same years, was to accomplish the almost impossible task of adding further lustre to his father's name, expecially in optics and astronomy. His early papers to the Society were concerned with the effects of crystals on polarized light. During the next ten years Herschel was much occupied with affairs in London, but he returned in 1832 to give an account of Babbage's analytical engine. While Herschel was at the Cape of Good Hope, observing the Southern stars, letters from him to the Society were read on several occasions.

The third mathematician was William Whewell (1794–1866), who was to become an outstanding Cambridge figure as Master of Trinity; his learning and influence extended far beyond mathematics, as exemplified by his much read *History* and *Philosophy* of the natural sciences and his tenure (1838–55) of the Knightbridge Chair of Moral Philosophy. He was to be one of the chief reformers of the University and a consistent pillar of the British Association. In his earlier years, however, Whewell was a straightforward exponent of Leibnizian analysis and modern theoretical mechanics, though his first University post was that of Professor of Mineralogy, when Henslow released that Chair (1828). During the first decade of the Society's existence he presented no fewer than 23 papers ranging from geometry, optics, and mechanics to crystallography and mineralogy, with two on chemical notation and Gothic architecture for good measure; he was besides Secretary of the Society on several occasions (the last was 1836) and President 1843–45. In later life Whewell's vigour merged into authoritarianism; in Sidney Smith's phrase, science was his strength, omniscience his foible. He was a giant of mid-Victorian England, and a good modern life of him would be worth having.

Though he did not appear before the Society until 1823, with an historical paper on "The Analytical Discoveries of Newton and his Contemporaries", George Peacock (1791–1858) was closely allied to this same group; he was to hold all three of the Society's principal offices (President, 1841–43). Having taught mathematics with success at Trinity and published a celebrated text-book (*Algebra*, 1830) he was chosen for the Lowndean Chair in 1836, though astronomy was obviously not his subject. Perhaps partly in consequence he aroused some odium when, after his appointment as Dean of Ely (1839), he continued to hold the Chair without lecturing. He was more energetic in his attention to the fabric of his Cathedral.

William Whewell 1794–1866

Charles Babbage 1792–1871

George Peacock 1791–1858

There were many other mathematicians and physicists contributing occasionally in these early years, among them an honorary Scottish member, David Brewster (1781–1868), another great figure in the British Association and in scientific education, but Sedgwick, Henslow and others were hardly less active in discussions of geology, minerals, meteorology, pathology, and a surprising number of technological topics.

The nineteenth century was also a period of great development in observational astronomy—previously neglected in Cambridge apart from the efforts of Thomas Catton (1760–1838) at the St. John's College Observatory; in this the University began to play a part after the opening of its Observatory near Madingley Road in 1823, at a cost of £19,000. George Biddell Airy (1801–92) was elected Plumian Professor and given its direction in 1828. His first paper—on silvered telescope mirrors, one of the great technical steps—was read by Peacock to the Philosophical Society when Airy was still a Scholar of Trinity (November, 1822). This was the first of a long series, in which incidentally Airy described for the first time (1825) the visual defect, astigmatism, which he had discerned in his own eyes, and its correction. After his removal to Greenwich in 1835 Airy became one of the greatest Astronomers Royal. Airy, like his successor in Cambridge astronomy James Challis (1803–82), figured unfortunately in connection with the work of John Couch Adams; Challis was the Society's President at the critical time (1845–47). Director of the Cambridge Observatory for twenty-five years after 1836, and Plumian Professor from the same date to the end of his life, if Challis did not win great fame it was not through lack of vigour. He was a prolific author, devoting the later years of his life to relating the known laws of gravity, light, heat, electromagnetism etc. to the action of a single aether. He lectured steadily on practical astronomy from 1843 to 1879, when the course was printed, but the papers he had published by the Society were mostly on fluid mechanics. Unfortunately, he was less active in following up Adams's indications of the position of a hypothetical new planet, which might be disturbing the motions of Uranus. As a result (in his own words) 'I lost the opportunity of announcing the discovery, by deferring the discussion of the observations, being much occupied with reductions of comet observations, and little suspecting that the indications of theory were accurate enough to give a chance of discovery in so short a time.' In fact neither

Sir George Biddell Airy 1801–1892   James Challis 1803–1882

Challis at Cambridge nor Airy at Greenwich paid much attention to the indications for a search given them by Adams in September, 1845, until the following July when they read papers by U. J. J. Leverrier who had independently reached—and more wisely, published—a similar prediction. Challis began elaborately to plot all the stars in the likely area of the sky—but in fact missed the new planet, though it was later ascertained that he had twice recorded Neptune in his registers, without realizing that this was the object he was seeking. Galle at Berlin, primed by a letter from Leverrier (September, 1846), within an hour spotted a body not shown in Bremiker's atlas. This on subsequent examination proved to be Neptune.

John Couch Adams (1819–92), who was thus disappointed, went on to become Lowndean Professor (1858) and in succession to Challis Director of the Observatory (1861). He had the

15

John Couch Adams 1819–1892

curious double distinction of refusing a knighthood, and having a University Prize founded under his name during his own lifetime, when the ingenuity of his solution to the inverse problem of perturbations had won the esteem it merited.* He presided over the Philosophical Society from 1861 to 1863. Before this, however, in 1853, Adams had made another important contribution by showing that the Sun's part in the Moon's secular acceleration could not amount to more than $5 \cdot 7''$ per century, about half the observed value; much later the phenomenon of

---

* It should be noted, however, that the real characteristics of Neptune's orbit differ markedly from those predicted by both Leverrier and Adams. Both had relied on 'Bode's Law' as a guide to Neptune's mean distance from the Sun—Challis wrote a paper in the *Transactions* (III, 1830) on the extension of 'Bode's Law' to planetary satellites—but Neptune's orbit falsifies 'Bode's Law'.

tidal friction was invoked to account for the remainder. Rather similarly Adams's proof (1867) that the Leonid meteor shower has a period of about 33 years enabled the hypothesis to be entertained that such meteors represent the debris of a former comet (*Proceedings*, 2, 1867).

Curiously enough, one of the very few first-hand descriptions of a meeting of the Society at this early period relates to that of 18 March 1867 at which, after a short communication from Challis, Adams described this last piece of work. The writer was W. P. Turnbull, father of the mathematician H. W. Turnbull, and his essay was written for a club called the Parallelepiped:

'It was gaslight. The general audience—among these were Professors Cayley and Miller, and a few ladies who would lend countenance to philosophy—were seated on tiers of benches, and below this general audience sat an audience fewer and more select; a more vital part of the Society, I suppose, whether by merit or by office. . . . Behind these worthies there is an array of large diagrams intended by Adams to illustrate his discourse. Challis also has something written out on a blackboard, and both the lecturers are accompanied (silently) by instruments. Distance and the careful closing of doors and windows keep out the noise and air of the world.'

It was, apparently, an apologetic evening. Challis apologised for being less interesting than Adams, and indeed his communication was an exercise in triangulation whereby he discovered the difference in longitude between the Society's clock and that in the Observatory. Adams apologised for giving his paper a mathematical rather than a more enticing form. He explained the five possible hypotheses by which the periodic occurrence of the meteor-shower could be explained, and his reasons for preferring that attributing a long elliptical orbit and a period of about 33 years to the mass of debris. In the discussion Challis was critical of Adams's preference. Our undergraduate witness's comments make it plain that the events of twenty years earlier were by no means forgotten. 'I prefer Adams's authority on this point,' he writes, (somewhat illogically) 'because all men judge that Adams's prediction with regard to Neptune was correct; and I remember against Challis that he was not altogether propitious to the efforts of Adams on the former occasion. Had it not been for Challis, I suspect that England, and not France, would have had the glory (slight as it was) of priority in that discovery.'

Peacock, Whewell and Challis were three successive Presidents distinguished as mathematicians, and so indeed was their successor, Henry Philpott (1807–92), Master of St. Catharine's, since he had been Senior Wrangler, but he became better known as an ecclesiastic, as Prince Albert's man in the University, and as an agent of University reform in the 1870s. After him followed (in the years 1849–51) Robert Willis (1800–75), a most extraordinary man, who had already served the Philosophical Society three times as Secretary. His basic interest was mechanical, but he expressed it in a somewhat odd way—after his election to a Fellowship at Caius—by an investigation of the mechanism of voice-production, on which he contributed papers to the *Transactions* (1828–29). In 1837 he was elected Jacksonian Professor, devoting himself to the perfection of Farish's system of interchangeable mechanical parts and the general theory of machines (*Principles of Mechanism*, 1841). Meanwhile he had plunged with tremendous enthusiasm and energy into the study of medieval architecture, writing one of the first books on the 'Gothic' style;* here his analytical abilities were as well displayed as they were in engineering. Between 1846 and 1864 he published a stream of monographs on the architectural history of the English cathedrals, and he began an elaborate study of the buildings of Cambridge that was completed by his colleague John Willis Clark (President, 1888–90).

The tradition of Henslow was continued in Charles Cardale Babington (1808–95), a botanist who served the Society many times as Secretary between 1851 and 1866. The fact that he had no field experience outside Britain, and that his treatment of his subject was antiquated, did not prevent his *Manual of British Botany* (1843 etc.) being a most successful textbook. In choosing as President in 1851 William Hopkins (1793–1866)—who was selected by the Geological Society for the same office in the same year—the Society honoured a very different type of man, who became one of the great Cambridge 'coaches' even though he himself only began his studies at Peterhouse in his thirtieth year. George Stokes, William Thomson and James Clerk Maxwell were among his pupils—seventeen Senior Wranglers in all. It was Sedgwick's stimulus that sent him to 'physical geology' as the chief subject of his own researches, on which he contri-

---

* He was perhaps stimulated to this by Whewell, who gave talks on this subject to the Society in 1832–33.

William Hopkins 1793–1866   C. C. Babington 1808–1895

buted a number of papers to the Society's publications. After his death the Society perpetuated his memory by founding a triennially awarded prize bearing his name.

After Sedgwick's second term (1853–55; he and Sir George Darwin were unique in enjoying this distinction) the Presidency passed to Sir George Edward Paget (1809–92),* long the medical director of Addenbrooke's Hospital and the first doctor to examine candidates for degrees at the bedside, and then to William Hallowes Miller (1801–80), Professor of Mineralogy, who had contributed notably to crystallography (*Transactions*, III, 1830); and who later worked on metrology—after the destruction by fire of the national standards in 1834—and also on optics (*Transactions*, VII; *Proceedings*, 2). During his second year of office there occurred an event of an unusual kind in the history of the Cambridge Philosophical Society: Adam Sedgwick's vehement and public attack on the concept of biological evolution as put forward by Charles Darwin in the previous year (7 May 1860). Although papers from Darwin were read before the Society in 1835 and 1837 his connection with it was never close—he was never a Fellow—and he did not find the

---

* As in the days of the notorious resurrection-men, all teachers of anatomy suffered from a shortage of human material. On one occasion Paget received a hamper containing the body of the victim of a shooting-accident from a sporting friend who used to supply him with game. The sternum, full of shot, is still in the Museum of Pathology.

prospect of Sedgwick's criticism particularly alarming. This was indeed powerfully expressed, but hardly novel. Sedgwick argued (and for the present it was true) that there was no palaeontological evidence in favour of evolution: in the transitions from the Carboniferous to the Permian, from Permian to the 'Muschel Kalk', and from this to the Lias, although successive 'creations' of flora and fauna succeeded each other, it was impossible 'to find the connecting organic links which bind together the organic types of the two periods . . . in our step onward we discover no point of rest for the hypothesis which derives the newer types by natural generation from the older.' Furthermore, Sedgwick alleged from the phenomena of the 'mutual adaptation of parts' and the 'wonderful adaptation of organs to all the complicated conditions of the surrounding world' the 'existence of a great, wise and prescient First Cause', and argued that by rejecting final causes the hypothesis of evolution promoted 'the first step towards atheism'. To reinforce this point and conclude his lecture Sedgwick

> read a variety of extracts from modern works which appealed to Darwin's hypothesis for proof; and then utterly repudiated all that was supernatural in Gospel history, making the New Testament no better than a myth, and its miracles nothing better than a series of falsehoods, or delusions, or ignorant impostures.*

The relevance of the quotations to a scientific debate is not clear, since Sedgwick admitted that Darwin was not responsible for these conclusions; it seems (from a letter of Darwin's to Hooker) that Henslow defended Darwin, presumably against this *odium theologicum*. However, Darwin refused to consider Sedgwick's criticisms and aspersions important. 'As for the old fogies at Cambridge', he wrote to Hooker, 'it really signifies nothing. I look on their attacks as a proof that our own work is worth the doing.' What a change since 1831 when Darwin had been so impressed by Sedgwick's grasp of the manner in which science groups facts 'so that general laws or conclusions may be drawn from them.' (*Autobiography*)!

It is a pity that there was no Huxley, Mendel nor Marsh in the Cambridge Philosophical Society to promote the vigorous development of Darwin's discoveries.

---

* The report is taken from the *Cambridge Herald and Huntingdonshire Gazette*, 13 May 1860. The Rede Lecturer (John Phillips, Professor of Geology at Oxford) took a similar line about the same time.

A few words should be added about some notable Fellows and contributors to the Society's journals who were less prominent in its business or the public eye. One of them, Augustus De Morgan (1806–71), was Professor of Mathematics at University College, London, from 1828 to 1831 and again from 1836 to 1866; at Cambridge he had been a close friend and associate of Whewell, Herschel and Peacock, and so continued—though unlike most Cambridge men he detested the Mathematical Tripos and indeed every form of competitive examination. De Morgan was decidedly eccentric on a number of points. He published many papers on pure mathematics and logic in the *Transactions* and *Proceedings* between 1833 and 1869; he was a collector of old mathematical books, and did a good deal of useful and unconventional work on the history of mathematics and science.

An even stranger and far more creative figure than De Morgan was George Green (1793–1841), a self-taught mathematical virtuoso whose abilities were wasted in a tragedy of frustration, arising from the limited horizons of English provincial life. Green's father was the owner of a windmill at Sneinton, Nottingham, where the son assisted in running his milling and baking business. Mathematics he pursued in his spare time in the windmill, studying almost unaided the work of the French school of mathematicians especially Poisson and Fourier. In 1828 he published at Nottingham with the aid of a few private subscribers *An Essay on the Application of Mathematical Analysis to the Theories of Electricity and Magnetism*, in which he introduced the formula relating surface and volume integrals (since always known as Green's Theorem) and applied it to the concept of potential, also original with him. Although the book was generally ignored, it brought George Green to the notice of Sir Edward Bromhead, a Fellow of the Cambridge Philosophical Society, who encouraged him to prepare two papers published in the Society's *Transactions* (Vol. V, 1833 and 1835) and a third for the *Transactions* of the Royal Society of Edinburgh. Further, he persuaded Green (who had now sold the family business) to go to Cambridge, where he was placed Fourth Wrangler in the Tripos of 1837, at the age of forty-four. In the same year he was elected a Fellow of the Philosophical Society (to which he contributed six more papers) and soon after was appointed to a Fellowship at Caius. But just as he was winning a place, his health failed completely. That George Green had not succeeded even now in making his scientific work well-known, despite the Philosophical Society's support, appears from a letter of Lord Kelvin to Sir Joseph Larmor written long afterwards:

> When I went up to Cambridge as a freshman, I asked at all the book shops in Cambridge for Green's Essay on Electricity and Magnetism, and could hear nothing of it.
>
> The day before I left Cambridge for Paris after taking my degree, in Jan. 1845, I met Hopkins on what I believe was then called the Senior Wrangler's walk, and I told him I had enquired in vain for Green's Essay and had never been able to learn anything about it all the time I was an undergraduate. He said "I have some copies of it." He turned with me and took me to his house, and there, in his chief coaching room in which I had been day after day for two years, he found three copies of Green's Essay in his bookcase and gave them to me.
>
> I had only time that evening to look at some pages in it, which astonished me. Next day, if I remember right, on the top of a diligence on my way to Paris, I managed to read some more of it.

Kelvin was so enthusiastic that he made several scientific men in Paris read Green's *Essay*, which he caused to be reprinted in the following year, 1846. In modern times Green's all too brief association with the University and its Philosophical Society has been paralleled by that of Srinivasa Ramanujan (1887–1920), the Indian mathematician who was brought to Cambridge by G. H. Hardy in 1914 from his work as a clerk in the port of Madras. There he enjoyed just five years of creative activity.

Finally, three minor figures may be recorded here, mainly for their appearance in the Society's publications since their careers took them away from Cambridge. Firstly there is the respectable geologist D. T. Ansted (1814–80), professor at King's College, London, formerly a Fellow of Jesus. Secondly, Phillip Kelland (1808–79), who, after being Senior Wrangler and first Smith's prizeman in 1834, had a distinguished career as Professor of Mathematics at Edinburgh from 1838 onwards. He was President of the Royal Society of Edinburgh in the year before his death. Kelland's early work printed in the *Transactións* from 1836 to 1841—his connection with the Society was brief—was largely concerned with physical optics, a science attracting much interest after the discoveries of Fresnel had been

Srinivasa Ramanujan 1887–1920

assimilated. Thirdly, practical optics attracted some attention from Richard Potter (1799–1886), though most of the papers he sent to the Society were on mechanics. Potter, though once a pupil of John Dalton, was another who came to science late in life, becoming a Fellow of Queens' in 1839. From 1841 to 1865 he was a colleague of De Morgan's at University College, London; the last years of his life were spent as a resident in Cambridge. These careers, together with those of much more brilliant men like Clerk Maxwell and William Thomson, cause one to wonder whether Cambridge at this time did not offer too small a measure of hospitality to the great volume of scientific talent among its students, of which the Mathematical Tripos was at least a partially accurate measure. Colleges could accommodate few such men, and the occurrence of vacant Fellowships was fortuitous; the University could provide for even fewer. Certainly the Philosophical Society would have been a far stronger body had the University been able to find room for a larger proportion of its scientifically able men to stay on for teaching and research.

CHAPTER THREE

The Society's Affairs
after 1819

THE FIRST meetings of the newly organised Philosophical Society were held in the Museum of the (old) Botanic Garden, built in 1784 on much the same site as the Austin Wing of the Cavendish Laboratory today; only in April 1820 was a move made to a 'commodious house'—rented rooms above the shop managed by one Bulstrode in Sidney Street, facing up Jesus Lane. In 1832 renewal of the lease was denied and the Society (urged by Peacock) resolved to lease a plot of land from St. John's College in All Saints' Passage, and build its own house, hoping that by this step 'the immediate convenience, the future welfare, and the permanent existence of the Society' would be secured. Money was raised by selling £50 bonds to Fellows, the plans drawn up by Mr. C. Humfrey (a Cambridge builder) were approved on 16 May 1832, and the Society was able to enjoy its new home from the autumn of 1833. This is the building until recently occupied by the Hawks Club, occasioning Hutchinson's quip that 'where Philosophers once studied the scientific journals athletes now turn over the pages of the Sporting Times.'

The manner in which the Philosophical Society made use of this building now occupied by the British Council, is not at all clear and one cannot tell whether meetings were held there or whether the Society continued to assemble for papers and discussion in lecture-rooms, as before and since. There was a carpeted Reading Room and a Library (*see below*) as well as apartments for the Curator, Mr Crouch. The house was from the beginning closed on Sundays, a rule reinforced by a prohibition (after 1843) of the use of lights or fire in the building on Sundays, except in the Curator's private rooms. In 1844 a group of Sabbatarian Fellows urged even more stringent rules against Sunday reading, but the Council felt it had gone far enough,

The House in All Saints' Passage built by the Society in 1832.

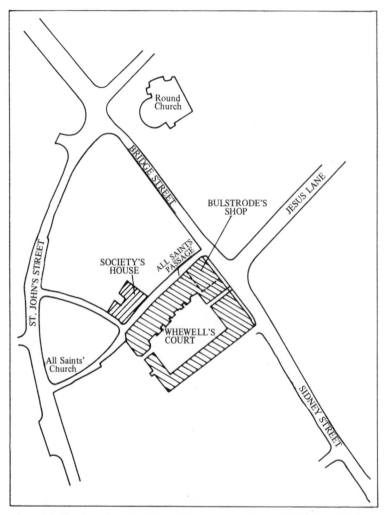

Plan of Bulstrode's shop (1820) and the Society's house (1832) in relation to present day buildings.

Not enough (or too much) to keep the wretched Crouch on the path of honesty, however; he was discovered to have robbed the Society of considerable sums.

The House also contained the Society's bulky collections, which by 1838 already constituted 'a valuable Museum of British and Foreign Natural History which contains extensive collections of British birds, shells, fish and insects and also a great number of fish from China, the coasts of S. America (col-

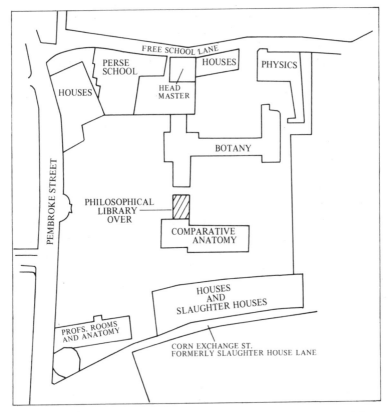

The Museums on the Site of the Old Botanic Garden, 1873. (Adapted from A. E. Shipley, *"J"*, *a memoir of John Willis Clark*).

lected and presented by C. Darwin Esq.) and other countries, besides numerous foreign insects and various curiosities', formed by purchase as well as gift. It was the intention of the Society's naturalists to assemble a full collection of the fauna of Cambridgeshire. Apart from the several hundred species of fish presented by Darwin, however, the natural history museum never really attained scientific importance.

For reasons not at all obvious membership of the Society failed to increase after the late 1840s (p. 67 below) and the House, never economic, proved an impossible expense. The Society was at no time affluent: in 1841–42, for example, it had an income (including composition fees and Reading Room subscriptions) of under £530, while expenses amounted to about £520. Hence there was no way of paying off the debt of over £2000 incurred in building the House, or of raising a sum for renewing the forty year lease from St. John's; moreover, the printing of the *Transactions* was delayed by lack of money. The Natural History collections were another source of expense, since a Curator was employed, though this was met by the generosity of a group of Fellows. Accordingly, when the University resolved in 1865 to build some new museum and lecture rooms the Society agreed (27 February) to ask for accommodation for its Library in the new buildings, and somewhat later offered the University as a gift the whole of its own Natural History collections, which were incorporated in the Museum of the new Department of Zoology and Comparative Anatomy.

This Museum was in the north-south range of the new buildings (*see the plan*), with the Philosophical Library housed near by in a room on the upper floor of the east-west range, just east of the archway passing through it. The first meeting of the Council in this building was held on 16 October 1865, and ordinary meetings began in a nearby lecture-room a few days later. The House in All Saints' Passage was put up to auction and sold for £1200, after an offer by the Pitt Club to rent it had been declined. To ease the Society's financial position a number of the Fellows simply tore up their bonds.

Thus began the formal association of the Society with the University, through the latter's recognition that the Society performed important functions in the intellectual life of Cambridge that the University itself did not undertake. The chief of these was undoubtedly the maintenance of the Philosophical Library, for there was nothing else comparable in the University and this alone enabled Cambridge scientists to follow developments on the continent and in the U.S.A. As science played an ever greater part in University studies, so the importance of the Library increased. Both the Library and the natural history collections had begun to grow as soon as the Society occupied its own apartments; as early as May 1820 Henslow had presented a number of specimens (for which cabinets had to be procured) while other Fellows presented books. In 1821 it was agreed to set up a Reading Room, where Fellows could find not only scientific journals but newspapers, literary reviews and the quarterlies—reading matter then not so readily accessible in Colleges as it is now, especially since in some the Junior Fellows were not privileged to enter the Combination Room, except on Feast Days. Fellows of the Society wishing to use the Reading

Room paid a separate subscription of one guinea. It was given up in 1856, when membership of the Society was declining. In building up the Society's library (always restricted as much as possible to periodicals rather than monographs) exchanges with other Societies have always been the main resource, little having been spent at any time on cash subscriptions. Indeed, a major part of the Council's business throughout most of its history has been deciding which exchanges to encourage and which to disallow. By about 1870 some 49 periodicals were obtained by exchange with foreign academies and societies, five of them being in the U.S.A. In addition, Poggendorf's *Annalen* and the *Annali Mathematici* were secured by exchanges, and thirty-five copies of the Society's own publications were sent to other institutions in Great Britain from whom in many cases a reciprocal favour was received. A list of the foreign exchanges in 1870, giving a view of principal scientific centres at this period, is appended to this chapter. The number of periodicals exchanged at the present time is about 1100.

By 1880 the space allowed had become too cramped so that the Society's 5,000 volumes made it almost unusable. Once again an appeal was made to the University, as better space was known to be available, Alfred Newton, President, writing to the Vice-Chancellor as follows (October, 1880):

'The Cambridge Philosophical Society occupies as you are aware a room in the New Museums which is used for the meetings of the Council of our Society and also as a Library. The books have now become so numerous that a larger apartment is required for their convenient accommodation.

The want of a central scientific library in the New Museums for the use of the Professors, Lecturers and students has long been felt, and recognized in various ways by the University. If such a library were founded and placed in a suitable room, the Council of the Philosophical Society would be prepared to recommend to the Society that their library should be deposited in it . . .'

Such a Science Library did not then and has not since come into existence; however, on the recommendation of a joint committee of the University and the Society the Senate passed a Grace granting the Society the use of a large room on the ground floor of the same building but to the west of the archway, so that Engineering was a near neighbour in the area later occupied by the Cavendish Laboratory. The University also

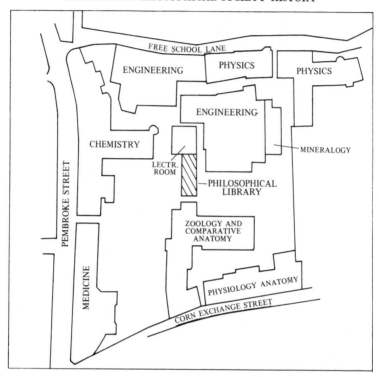

The Laboratories in 1912. (Adapted from A. E. Shipley, "*J*", *a memoir of John Willis Clark*).

undertook to pay the salary of a Librarian. In return the Society agreed, as Alfred Newton had proposed, that its Library should be opened for reference to all Professors, Curators and other teaching staff, reserving to Fellows of the Society the right to borrow books and periodicals and some other privileges. This right to borrow has remained one of the chief inducements for seeking election to the Fellowship. The management of the Library was transferred to a committee representing both the University and the Philosophical Society—though in fact the University representatives have often been Fellows too. In this way the Philosophical Library became, in effect, a valuable element in the University library system.

Many Fellows have made generous gifts of books and periodicals to the Library, among them John Willis Clark, Sir G. M. Humphry, Francis Darwin, and Sir Joseph Larmor. However, many books have been dispersed over the years by transfer to University Departments, loan or sale, partly in order to main-

The Library 1969.

tain the Library's scientific and periodical character. The great strength of the Library in scientific periodicals was recognised when it was renamed officially (in 1967) the Scientific Periodicals Library.

A small portion of the building in which the Philosophical Library was housed from 1881 to 1935, and in which meetings were held during the same period, still stands on the New Museums site looking fussily incongruous amid its plainer modern neighbours (*see the plan of 1912*). Suggestions were made for enlarging it, but these were never put into effect. With some minor adjustments the Philosophical Society was well enough content with its accommodation until, in the early 1930s, shelf-space became increasingly limited. Moreover, with the prospect of the new University Library (opened in 1935) coming into use and of other building improvements, the University was forced to reconsider its arrangements. In November 1933 a report of the Council of the Senate recommended that the Philosophical Library should be transferred to the upper rooms of the Arts School (built in 1910), providing an ample if inconvenient book-stack in the basement. The Council of the Philosophical Society variously preferred (on different occasions) either a redevelopment on the existing site, or a move to the then Squire Law Library on the other side of Downing Street (later the Marshall Economics Library). This would have been more convenient to many of the new biological laboratories. But it was to the Arts School that the Library moved in 1935.

Before the building of the house in All Saints' Passage was begun a decision was taken that the Society should acquire proper legal status by incorporation under a Royal Charter, which was granted by William IV. The King's cousin and Chancellor of the University, William Frederick, Duke of Gloucester, had been the Society's Patron since its foundation: his younger brother, Augustus Frederick, Duke of Sussex (and later President of the Royal Society) had been Vice-Patron since 1821. The Prince Consort, when elected Chancellor of the University (1847) also agreed to serve as Royal Patron of the Society and in July of that year was thanked for a 'munificent donation' to the Society's funds. The grant of the Charter in 1832 was made an occasion for a dinner, or 'blow-out' in Sedgwick's* phrase, which was often repeated on the anniversary, 6 November. This pleasant practice was renounced during the war years, 1914–18, but there was a festive celebration of the Society's centenary in the Hall of Sidney Sussex College (of which the President, C. T. R. Wilson, was a Fellow) in 1919. Old Professor Liveing was among those present, aged ninety-one. There were nine courses and ten speeches. Biennial dinners were resumed until 1932, in which year the grant of the Charter was duly celebrated, Prince George and the Prime Minister being among the guests. Thereafter there were only two more dinners (in 1935 and 1938) until that arranged in 1966 to celebrate the visit to Cambridge of Academician and Mrs. P. Kapitza.

The founders had from the first intended that the Society should be more than a forum for the presentation and discussion of papers. Their preservation was early assured by the commencement of the *Transactions* in 1821; twenty-three volumes had been completed (at irregular intervals) when this series was terminated in 1928.† The current *Proceedings* first appeared in 1844, and since 1928 quarterly parts have been published regularly. Originally its contents were precisely as its name suggests, as it contained all communications to the Society either in abstract, or in full. Some briefly abstracted papers were reserved for full publication in the *Transactions*. A committee of

---

* Adam Sedgwick's was (for reasons of economy) the only name to appear in the Charter as a Fellow of the Philosophical Society.

† An Author Index of the *Transactions* and the *Proceedings* Vols 1–50 was published in 1961, and *Proceedings* is now indexed every ten years.

The Charter

WILLIAM the Fourth

£30

of the United Kingdom of Great Britain and Ireland King Defender of the Faith To all to w...
Fellow of Trinity College and Woodwardian Professor in the University of Cambridge has by his Peti...
did in the Year One thousand eight hundred and nineteen form themselves into a Society for the promotion
have also collected and become possessed of a valuable Library and various Collections in Natural History
Philosophical and Natural Knowledge among the Graduates of the said University by offering encourage...
read before the Society And whereas the said Petitioner believing that the well being and usefuln...
of himself and other Fellows of the said Society most humbly prayed that We would be pleased to grant a Roy...
become Fellows thereof subject to such Regulations and Restrictions as to Us may seem good and expedient
mere motion Willed granted and ordained and Do by these Presents for Us our heirs and Successors will gran...
are now Fellows of the said Society or who shall at any time hereafter become Fellows thereof according to such
Name and Style of the Cambridge Philosophical Society by which name and style they shall have p...
the same name and style to sue and be sued to implead and be impleaded to answer and be answered in...
enjoy to them and their Successors any Goods and Chattels whatsoever and also be able one capable in...
a Hall or College and any messuages lands tenements or hereditaments whatsoever the yearly value of
the same respectively at the Rack Rent which might have been had or gotten for the same respectively at t...
as fully and effectually to all intents effects constructions and purposes whatsoever as any other of our
any disability might or could act or do in their respective concerns And We do hereby grant our co...
grant sell alien and convey in Mortmain unto and to the use of the said Society and their Successors any
and We further grant and declare that there shall be a General Meeting of the Fellows of the said Body Politi...
and manage the concerns of the said Body Politic and Corporate And our Will and Pleasure is that our
and our most dearly beloved Brother Augustus Frederick Duke of Sussex our right trusty and right...
Chancellor of the said University shall be the Vice Patrons of the said Body Politic and Corporate And
not more than three Secretaries and not more than twelve or less than seven other Fellows to be elected...
six Calendar Months after the date of this our Charter and that the said Adam Sedgwick shall be the fir...
hereby further Will grant and declare that it shall be lawful for the Fellows of the said Body Politic and Corpor...
That the General Meetings shall choose a President Vice Presidents Treasurer Secretaries and other Members
the regulation of the said Body Politic and Corporate for fixing the days on which the Ordinary Meetings of the Society...
of the Estates Goods and Affairs of the said Body Politic and Corporate and for fixing and determining the manner of...
And We do further declare that it shall be lawful at the General Meetings of the said Body Politic and Corporate to...
Pleasure that no Fellow who has filled the office of President for two successive years shall be again eligible to the sa...
Meetings as aforesaid shall without the consent of the Council of the said Society or Body Politic have the power of altering or
desirous of so altering or repealing any such existing Bye Law or of making any new one shall have given to the Council of the...
Council shall but not inconsistently with or contrary to the provisions of this our Charter or any existing Bye Law or the...
the said Body Politic and Corporate and also the entire management and superintendence of all the other Affairs and Concerns r...
the objects and views of the said Body Politic and Corporate And We further Will grant and declare that the whole Prop...
shall have full power and authority to sell alienate change or otherwise dispose of the same as they shall think proper but...
Politic or Corporate shall be made except with the approbation and concurrence of a General Meeting And Wh... has...
the said Body Politic and Corporate in opposition to the general scope true intent and meaning of this our Charter or...
shall be made the same shall be absolutely null and void to all intents effects constructions and purposes whatso...
Westminster this Sixth Day of August in the third Year of our Reign.

By Writ of

by the Grace of God

...sents shall come Greeting Whereas Adam Sedgwick Clerk Master of Arts ...ented unto Us That he together with others of our loyal Subjects Graduates of the said University ...Natural History and have subscribed and collected considerable Sums of money for such purposes and ...to a considerable amount and have also been and continue to be actively employed in the Promotion of ...esearch and especially by the Publication of Volumes of Transactions composed of Original Memoirs ...y would be most materially promoted by obtaining a Charter of Incorporation hath therefore on behalf ...porating into a Society the several persons who have already become Fellows or who may at any time hereafter ...at We being desirous of encouraging a design so laudable have of our especial grace certain knowledge and ...he said Adam Sedgwick Clerk and such others of our loving Subjects as have formed themselves into and ...hereafter be formed or created shall by virtue of these Presents be one Body Politic and Corporate by the ...and a Common Seal with full power to alter vary break and renew the same at their discretion and by ...f Us our heirs and successors and be for ever capable in the Law to purchase receive hold possess and ...standing the Statutes of Mortmain) to take purchase possess hold and enjoy to them and their Successors ...the said Hall or College shall not exceed in the whole the Sum of Two thousand Pounds computing ...hase or Acquisition thereof and to act and do in all things relating to the said Body Politic and Corporate ...any other Body Politic or Corporate in our United Kingdom of Great Britain and Ireland (not being under ...Authority into all and every person and persons Bodies politic and Corporate (otherwise competent) to ...tements or hereditaments not exceeding such annual value as aforesaid And our Will and Pleasure is ...e held from time to time as hereinafter mentioned and that there shall always be a Council to direct ...o Cousin William Frederick Duke of Gloucester Chancellor of the said University shall be the Patron ...and Councillor Philip Earl of Hardwicke the High Steward of the said University and the Vice- ...will grant and declare that the Council shall consist of a President three Vice Presidents one Treasurer ...f the said Body Politic and Corporate and that the first Members of the Council shall be elected within ...said Body Politic and Corporate and shall continue such until the Election as aforesaid And we do ...ed to hold General Meetings once in the year or oftener for the purposes hereinafter mentioned viz! ...the General Meetings shall make and establish such Bye-Laws as they shall deem to be useful and necessary for ...aining the mode in which Fellows and Honorary Members shall be elected admitted or expelled for the management ...ice Presidents Treasurer Secretaries and other Members of the Council and the period of their continuance in office ...er Bye Laws and to make such new Bye Laws as they shall think useful and expedient But it is our Will and ...e expiration of one year from the termination of his office And our Will and Pleasure is that no such General ...ing Bye Law or of making any new one unless the Fellows or Fellows of the said Society or Body Politic who shall be ...Politic one months previous notice of such his or their intention And we further Will grant and declare that the ...f this our Realm or the Statutes of the University of Cambridge have the sole management of the Income and funds of ...ay do all such acts and deeds as shall appear to them necessary or essential to be done for the purpose of carrying into effect ...Politic and Corporate shall be vested and we do hereby vest the same solely and absolutely in the Fellows thereof and that they ...age Incumbrance or other disposition of any messuage lands tenements or hereditaments belonging to the said Body ...our Royal Will and Pleasure that no Resolution or Bye Law shall on any account or pretence whatsoever be made by ...ates of this Realm or the Statutes of the University of Cambridge and that if any such Resolution or Bye Law ...ess whereof We have caused these our Letters to be made Patent Witness ourself at our Palace at

...al

Brougham

TRANSACTIONS

OF THE

CAMBRIDGE

PHILOSOPHICAL SOCIETY.

VOL. I.   PART I.

CAMBRIDGE:

PRINTED AT THE UNIVERSITY PRESS,

AND SOLD BY

DEIGHTON & SONS, AND NICHOLSON & SON, CAMBRIDGE;
AND T. CADELL, STRAND, LONDON.

M.DCCC.XXI.

Title page of the first issue of the "Transactions".

the Council in 1876 recommended that there should be two issues per year, that the Secretaries' report should be printed, and that the Secretaries might also accept short papers for publication in the *Proceedings*. With the increasing professionalism of science in the early twentieth century, and with more specialist scientific journals continually appearing, the position of the Society's publications became less happy, with regard both to their reputation and the quality of material offered for publication. The 1914–18 war did not help matters. The Great

War (as it was once called) leaves surprisingly few traces in the Society's history: there were difficulties about obtaining German periodicals, the provision of tea at meetings ceased, and notices were received from Whitehall urging caution in the publication of scientific matters that might affect the war effort. But ultimately through inflation the war did the Society great harm.

The first proposals for a reform in the Society's publications were made in 1917 by G. H. Hardy, who was concerned about the low standard of papers in pure mathematics coming from the Society. By this time the *Proceedings* were almost wholly devoted to topics in mathematics and physics, whereas the *Transactions* was a miscellaneous collection including work on biology also, in accord with the Society's traditional policy. Hardy proposed that the Philosophical Society, like the Royal Society, should publish simply *Proceedings A* and *B*, as well as making other reforms rendering *A* more attractive to mathematical authors. However, a committee preferred instead to recommend that, with the abandonment of the *Transactions* and the continued publication of the *Proceedings* unchanged, there should be founded (under the Society's direction) a *Cambridge Journal of Mathematics and Natural Philosophy* to be owned by the University Press. There was, naturally, opposition; a Special General Meeting of the Society was held to discuss publications on 12 May 1919, and it seems (for no record survives) that the proposed innovations were defeated.

At the same time pressure was felt from the biological Fellows who also regarded the existing journals as inadequate, despite *ad hoc* steps taken since the crisis in 1919. They too felt that the Society should support a journal devoted solely to their interests; hence, in 1922, the decision of the Society's Council to publish biological papers separately. *Proceedings of the Cambridge Philosophical Society, Biological Sciences*, accordingly first appeared in 1923. This was really a revival of Hardy's original idea, but put to the test it seemed that the supply of first-class papers was too small. It was (Sir) James Gray who suggested that review articles be commissioned for the new journal, the title *Biological Reviews and Biological Proceedings* being approved in December 1926, and being reduced to *Biological Reviews* from 1935 onwards. Virtually all the review articles in this quarterly were commissioned by the editor, H. Munro Fox, whose energy and knowledge continued to extend the new journal's success

Harold Munro Fox 1889–1967

long after his own removal to a Chair at Birmingham. With the commencement of *Biological Reviews* the Society began to emerge from its post-war doldrums into its recent period of prosperity, for the *Reviews* proved really useful to biologists and were greatly in demand. Through Munro Fox the Society made an important and lasting contribution to scientific communication and criticism.

A later contribution made by the Philosophical Society towards the encouragement of research in biology which may conveniently be noted here was its establishment in 1964 of a prize to be awarded in every third year to a member of Cam-

William Bate Hardy 1864–1934

bridge University who has done distinguished work in any
department of biology. Like the Hopkins Prize, its present value
is two hundred pounds. In naming this prize the Society com-
memorated Sir William Bate Hardy (1864–1934), a Fellow for
over 45 years, who carried out important investigations in such
varied fields as histology, food science and lubrication. He also
did much public work and was superintendent of the Low
Temperature Research Station in Cambridge from 1922 to the
year of his death.

But to return to the *Proceedings*: as Cambridge science
flourished in the 1920s and 1930s the number of communica-

tions offered to the Society far exceeded the amount of time available at meetings, so that it became necessary to 'read' many by title only, and the *Proceedings* became such in name alone, since many papers read were never printed, and many papers printed had never been read. It is interesting to note that scientists whose names are greatly respected today on occasion gave communications which the Philosophical Society, at any rate, did not print:

13 November 1911	Niels Bohr: 'The electron theory of metals'
28 February 1921	Sir Joseph Larmor: 'On the nature of crystal reflection by X-rays'
2 May 1921	Eric Rideal: 'On active molecules in physical and chemical reactions'
27 February 1922	E. V. Appleton: 'The direct detection of wireless waves'
3 December 1923	C. T. R. Wilson: 'A simple stereoscope'
25 November 1929	P. A. M. Dirac: 'A theory of electrons and protons'

And two examples from the same meeting on the biological side:

1 March 1926	Joseph Needham: 'Intracellular oxidation—reduction potential and anaerobiosis'
	James Gray: 'The growth of fish during the phase of yolk-sac nutrition'.

Nor were any of the evening lectures delivered during the nineteen-twenties and thirties printed. After 1945 there was again a change, whereby the *Proceedings* lost all but a nominal connection with the business at the Society's meetings; it is now a journal to which any suitable article may be submitted.

The Philosophical Society has also been responsible for a number of special publications, some of which will be mentioned later. One proposal that unfortunately came to nothing (it was a far greater task than anyone then imagined) originated with Sir Joseph Larmor in 1903, when he suggested that a new and complete edition of the works of Sir Isaac Newton should be put in hand, and in the following year the Council resolved to inform the University that the Philosophical Society contemplated 'a critical edition of Newton's scientific writings, together with the scattered correspondence and an annotated edition in English of the *Principia*.' R. A. Sampson (1866–1939) was to chair an editorial Board.* No more is heard until a Committee was again appointed to assist Professor Sampson in 1908; and that was the last of it, with the exception of another appointment of an advisory committee with the same object in 1924, the relevance of which is not explained. Larmor seems to have been the moving spirit, for he was interested enough, some years later, to print correspondence in which Maxwell described his early ideas on electricity (*Proceedings*, Vol. 32).

The latest special scientific publication of the Society has been the *Photometric Atlas of the Spectrum of Arcturus* (1968), prepared by R. F. Griffin of the Cambridge Observatories with the coudé spectrograph at the Mount Wilson Observatory using the 100-inch reflector. The spectrum, covering the wavelengths from 3600 to 8825 Ångström units, is spread over 374 tracings reproduced in facsimile. This is by far the most detailed stellar spectrum ever published so far, other than that of the Sun itself.

* * *

The Society has naturally participated in many public activities centred upon Cambridge, beginning with the third meeting of the British Association, in June 1833, over which Sedgwick presided. The Association met at Cambridge a second time in 1845, with Sir John Herschel as President and again in 1862; Sedgwick relates the manner of the invitation:

> Yesterday evening ... I was obliged to go to the meeting of the Philosophical Society. For I had to make a motion, on the part of the Society, to invite the British Association to Cambridge in 1862 ... we were of one mind; and I was happy to be the ostensible leader of the invitation, and to accept the leadership of the deputation.

This meeting was presided over by another Fellow of the Society, Robert Willis (1800–75), Jacksonian Professor since 1837, a distinguished engineer and architectual historian. The fourth visit of the Association to Cambridge was in 1904, when Arthur Balfour was President; the Philosophical Society arranged an 'At Home' in St. John's College. Later visits of the British Association to Cambridge took place in 1938 and 1965.

---

* Then Professor of Mathematics and Astronomy at Durham, later Astronomer Royal for Scotland.

Sir George Darwin 1845–1912

In 1908 a proposal made by the President, Professor E. W. Hobson, that the Society should join in the celebrations of the centenary of Charles Darwin's birth by publishing a volume of commemorative essays, was adopted and commended to the University Press, who agreed to publish the book. A committee including (among others) William Bateson, Francis Darwin, J. E. Marr and A. E. Shipley was put to work to call contributors together, Professor A. C. Seward being the editor. A stout volume reviewing Darwin's place in the history of the sciences, *Darwin and Modern Science*, duly appeared in the following year. The Society also played a part in other Cambridge tributes. A somewhat similar review of the Philosophical Society's own history was projected for its centennial year, 1919, but the times were unpropitious and the project fell through for lack of financial support.

While the Darwin celebrations were under way, the mathematician A. R. Forsyth proposed that, as the Fifth International Congress of Mathematicians was due to meet in 1912, and as the Congress had never assembled in England, it would be most appropriate to invite it to do so, and above all at Cambridge. The Society at once approved this proposal, an invitation was despatched to the Rome Congress in 1908, and duly accepted. The Fifth Congress accordingly was held in Cambridge during August 1912, Sir George Darwin, the President of the Philosophical Society, being chosen to preside over the Congress also. Among other activities a tribute was paid to the memory of Arthur Cayley (1821–95; President, 1869–71). A more unusual activity on the part of the Society was its organisation in 1936 of an exhibition of historic scientific instruments in Cambridge, for which the material was selected by Dr. R. T. Gunther.

PERIODICAL EXCHANGES, 1870

Amsterdam,	Akademie van Wetenschappen.
Batavia,	Bataviaasch Genootschap van Kunsten en Wetenschappen.
Berlin,	Akademie der Wissenschaften.
Bern,	Schweizerische Naturforschende Gesellschaft.
Bordeaux,	Société des Sciences.
Boston, U.S.	Boston Society of Natural History.
	American Academy of Arts and Sciences.
Brussels,	Académie Royale des Sciences.
,,	Observatoire.
Breslau,	Schlesische Gesellschaft.
Cadiz,	Observatory.
Calcutta,	Public Library.
,,	Geological Survey of India.
Cape of Good Hope,	Observatory.
Cherbourg,	Société Impériale des Sciences Naturelles.
Christiania [Oslo],	University.
Copenhagen,	Danske Videnskaberne Selskab.
Dantzig,	Naturforschende Gesellschaft.
Dresden,	K. Leopoldino-Carolinische Akademie.
Geneva,	Société de physique et d'Histoire Naturelle.

Göttingen,	Gesellschaft der Wissenschaften.
Kasan,	University.
Königsberg,	Physikalische Œkonomische Gesellschaft.
Leipzig,	K. Sächsische Gessellschaft der Wissenschaften.
,,	Astronomische Gesellschaft.
Lisbon,	Academia Real das Sciencias.
Lund,	University.
Melbourne,	Royal Society of Victoria.
,,	University.
Metz,	Académie des Sciences.
Milan,	R. Instituto Lombardo di Scienze e Lettere.
Moscow,	Société des Naturalistes.
Munich,	Bayerische Akademie der Wissenschaften.
,,	Sternwarte.
Paris,	L'Institut de France.
,,	Depôt de la Marine.
,,	Museum d'Histoire Naturelle.
,,	Société Géologique de France.
Philadelphia,	American Philosophical Society.
Pulkova,	Observatoire Impériale.
St. Petersburg,	Académie Impériale des Sciences.
,,	Observatoire Physique Centrale de Russie.
Sydney,	University.
Toronto,	Canadian Institute.
Upsala,	Royal Society of Sciences.
Vienna,	Akademie der Wissenschaften.
,,	K. K. Geologische Reichsanstalt.
Washington,	Smithsonian Institution.
,,	U.S. Naval Observatory.

CHAPTER FOUR

Victorian Heyday

W HEN THE *Origin of Species* was published the generation of Sedgwick, Henslow, Whewell and Babbage was settled in solid mid-Victorian middle-age; the radical fires of its youth were dead, and it seems that the Cambridge Philosophical Society had not been so successful as might have been hoped in giving a stimulus towards continued activity and reform. Although Cambridge in 1819 had stirred in its long slumber, it was not even by 1859 fully awakened and alert to the needs of a new age.

The Previous Examination ("Little-Go'), instituted in 1824 as an intermediate test of undergraduates' diligence, was a trivial affair (though it imposed extra papers in mathematics on Honours candidates). It was followed in 1848 by the Natural Sciences Tripos, upon which Honours could be awarded *if* the candidate had first qualified himself for a degree in arts, law or medicine. The examinations for this (and for the parallel Moral Sciences Tripos) were of a rather general character and it was clearly not intended that either of these new Triposes should rival in prestige the Mathematical Tripos or even the more recent Classical Tripos. By 1859 only 43 men had taken Honours in Natural Sciences. As matters of curriculum and examination were not dealt with by the Statutory Commissioners of 1856—through whom the government of the University was modernized—a reform of the Natural Sciences Tripos was effected after their work was done, in 1860. This among other things permitted a successful candidate with Honours to proceed to the B.A. degree. In 1875 more than twenty candidates were classed, and a few years before T. W. Danby had been awarded a Fellowship at Downing, mainly because of his performance in the Natural Sciences Tripos of 1864. Some—not all—Colleges began to appoint Lecturers in the various subjects of the

Tripos; possibly the most distinguished of such early appointments was that made to a Praelectorship in Physiology at Trinity College in 1870.

The new teacher, raised to the rank of Professor in 1883, was Michael Foster (1836–1907), who had been trained at University College, London, by William Sharpey (generally regarded as the first exponent of modern physiology in Britain) and also by Thomas Henry Huxley at South Kensington. It was on Huxley's teaching that Foster's own practical courses were modelled. He had many eminent pupils: among those most closely connected with the Philosophical Society were two future Presidents, Francis Maitland Balfour (1851–1882) and Francis Darwin (1848–1925). The former, an embryologist of great promise who fell an early victim to the passion for mountaineering in Switzerland, was elected to that office in the year preceding his death. Foster himself was elected for the years 1884–86. His own work on the heart is just reflected in the Society's *Proceedings* (Vol. 2), but he was best known for his *Textbook of Physiology* (1876) and his *History* of the same science (1901). Balfour is represented by several papers, as are also Foster's other students W. H. Gaskell (1847–1914)–one of the greatest of cardiac physiologists—and J. N. Langley (1852–1925) who took up Gaskell's work on the autonomic nervous system and succeeded to Foster's chair in 1903. From about 1890, however, the Philosophical Society ceased to reflect the success of the Cambridge school of physiology, partly because Foster and others had organised both a society and a journal for physiologists. Apart from scattered papers the work of Langley, W. M. Fletcher, Joseph Barcroft, F. G. Hopkins and C. S. Sherrington was to be published elsewhere. However, the Society did recognise the distinguished contributions to knowledge of the respiratory function of the blood and to biochemistry made respectively by Sir Joseph Barcroft (1872–1947) and Sir Frederick Gowland Hopkins (1861–1947) by electing both to the Presidency, the former in 1933, the latter in 1937. Both had studied with Foster.

At the time of Foster's appointment in 1870 the state of practical teaching in the sciences was very backward. St. John's College had equipped a chemical laboratory in 1853 and other Colleges gradually followed this example, indeed College laboratories remained in use until the 1920s. But it was always obvious that leadership in the teaching of science must come

Sir Michael Foster 1836–1907

F. M. Balfour 1851–1882

Sir Joseph Barcroft 1872–1947

Sir F. Gowland Hopkins 1861–1947

Front elevation of the New Lecture Rooms and Museums in the Old Botanic Garden. (From C. H. Cooper, *Memorials of Cambridge*, 1866).

from the University, and the provision of laboratories with it. In 1860 a fund was set up to provide new museums and lecture rooms on the former Botanic Garden site at the back of Corpus; these buildings were erected in 1865 and devoted to Botany and Comparative Anatomy with Zoology; here Foster began his work (*see plan* p. 27). As already mentioned, the Philosophical Library also was accommodated in these new buildings. A Chair of Zoology was created in 1866, the first holder being Alfred Newton (1829–1907), an ornithologist, whose visit to Spitzbergen two years before is described in the Society's *Proceedings*; this was his sole paper, though he had a long connection with the Society and was elected President in 1879. Soon a complex of laboratories occupied this site, as they have done (with many changing uses) ever since. Mineralogy was brought in, new rooms for Zoology built in 1876–79, an Engineering Workshop in 1874–84 and the rest of the Engineering school on the site of the Perse School in 1894–1900, the Medical Schools in 1900–1904, and a new Chemical Laboratory in 1907–08. Many of these buildings are still in daily use. But the most celebrated of all—also still in use—was the original nucleus of the Cavendish Laboratory, built in 1872–74, to which an extension was added in 1885. In 1868 a Syndicate had been appointed to consider the teaching of physics, especially heat, magnetism and electricity which, it was believed, should be treated experimentally and for which a new laboratory would be required. As the existing professors of mathematics and science declined to assume responsibility for these subjects it became clear that a new professor would be required too. The Chancellor, the Duke of Devonshire (whose family name was Cavendish), generously offered to provide the capital cost of the building and apparatus, leaving the University the task (which proved difficult) of finding money for the salaries of the Pro-

Alfred Newton 1829–1907   J. Clerk Maxwell 1831–1879

fessor and his Demonstrator—difficult (as Winstanley remarks) not because of hostility to Natural Science in Cambridge but because of the selfishness of the Colleges which were reluctant to contribute towards such new developments. James Clerk Maxwell (1831–79) was elected first Cavendish Professor of Experimental Physics in 1871, and President of the Philosophical Society in 1875—he and J. J. Thomson have been the only Cavendish Professors to hold this office so far, despite the Philosophical Society's close association with the thriving school of experimental physics in Cambridge.

Maxwell, who had with difficulty been persuaded to take this new post after already giving up his Chair at King's College, London, was of course a Cambridge man himself. He had been placed Second Wrangler in 1854,* in which year also he published his first paper in the Philosophical Society's *Proceedings* (Vol. 1) 'On the transformation of surfaces by bending'; among others of these early years is one (in the same volume) 'On Faraday's lines of force', which led Maxwell to the concept of the electromagnetic field. From 1856 to 1865 when he taught in Aberdeen and London his relationship with the Society was less active (though he remained a Fellow), but after his return he published eleven papers in the *Transactions* and *Proceedings*, in-

* The Senior being E. J. Routh, the great coach.

cluding one on the discoveries of the eighteenth century physicist Henry Cavendish to whom Maxwell devoted much attention in these last years. Another (*Transactions* XII, 1879) is on Boltzmann's equipartition theorem. J. J. Thomson has pointed out that (despite his papers on the electromagnetic theory) Maxwell was much less well known when he was appointed Cavendish Professor than he was after his famous *Treatise on Electricity and Magnetism* came out (1873). Nevertheless, the old dignitaries of the University were anxious to honour the new Professor; by accident or design his inaugural lecture was most obscurely announced, and the honours were perforce done at the first lecture of Maxwell's regular course. Sir Horace Lamb recalled: 'It was amusing to see the great mathematicians and philosophers of the place such as Adams, Cayley, Stokes, seated in the front row while Maxwell, with a perceptible twinkle in his eye, explained to them the difference between the Fahrenheit and Centigrade scales of temperature.' Maxwell rejoiced in a characteristically Victorian sense of humour, and it is not difficult to believe that this practical joke had been deliberately contrived.

At the time of his death, even though Maxwell had not succeeded in creating a great enthusiasm for physics among undergraduates, he had begun the creation of a post-graduate 'school' in the Cavendish; among those who worked under him were George Chrystal, later professor at Edinburgh; Donald MacAlister, later Principal of Glasgow University; Richard Glazebrook, later Director of the National Physical Laboratory; Arthur Schuster and Ambrose Fleming. The success of the rising school of physiology commenced at the same time, and there were Cambridge men distinguished in botany, geology and astronomy also, not to mention pure mathematics. Through the decade of the 1880s the Cambridge scientific scene was active as never before, and the list of names of men associated with the Philosophical Society who attained the highest ranks in science grows very long.

Maxwell's successor, the third Lord Rayleigh (1842–1919), contributed during his short tenure of the Cavendish Chair a number of papers to the Philosophical Society, one reflecting his great interest in acoustics (*Proceedings*, 4, 1881), and others his main concern while at the Cavendish, the determination of electrical standards. (It should be added that, as Professor, Rayleigh did a great deal to improve the teaching of physics and

Sir George Gabriel Stokes 1819–1903   Sir Richard Glazebrook 1854–1935

make the subject attractive to undergraduates). But a word should be said of the Philosophical Society's Presidents just recorded as attending Maxwell's first lecture. Sir George Stokes (1819–1903), one of the great English figures in mid-nineteenth century physics, in the centre of the classical tradition and a great friend of Kelvin's, published many papers in the Society's journals between 1842 and 1889; he was President 1859–61. The papers represent two of his main concerns: fluid dynamics and optics. Unlike Maxwell, Stokes never left Cambridge. He was elected into a Fellowship at Pembroke immediately after being placed Senior Wrangler in 1841—a Fellowship he voluntarily vacated by marriage in 1857, long before the change of Statutes, when he had been Lucasian Professor for some eight years. He served as one of the Secretaries of the Royal Society for over thirty years, was President 1885–90, and recipient of its Rumford and Copley Medals. Besides these offices he shared with Newton (with whom Stokes has often been compared) he, like Newton, served for several years as Member of Parliament for the University without ever speaking in the House. Stokes became a well known figure in Cambridge, because his kindness was as marked as his lack of dinner-table conversation. J. J.

Arthur Cayley 1821–1895

Thomson presents a delightful portrait of Stokes in his *Recollections and Reflections* (1936):

'The lectures I enjoyed the most were those by Sir George Stokes on Light. For clearness of exposition, beauty and aptness of the experiments, I have never heard their equal. He had only the simplest apparatus at his command, no light but that of the sun, no assistant to help him. He prepared the experiments before the lecture and performed them himself in the lecture, and they always came off.' On bright days, however, the audience might have to suffer privations because a noon lecture had been protracted well into the afternoon!

In 1899 the University celebrated Stokes's Jubilee year in the Lucasian professorship with pomp and real affection. Among other tributes he was presented with an elaborate address of congratulation by the Philosophical Society, together with a special eighteenth volume of the *Transactions* which was dedicated to him—and a very handsome volume it is.

The lectures of Arthur Cayley (1821–95) were on the other hand notoriously difficult, so much so that it is said that A.R. Forsyth (1858–1942), who introduced the theory of functions into Britain and was to be in the course of his long life an active participant in the affairs of the Philosophical Society, was Cayley's only real pupil; J. J. Thomson recalled that, when he went to hear Cayley, J. W. L. Glaisher and R. T. Wright (both M.A.s) sat on either side of the professor while Cayley wrote with a quill pen on large sheets of paper. Thomson, seated opposite, found note-taking somewhat difficult! Forsyth told a similar tale. Cayley's immensely productive career as a mathematician was not easy, for his refusal to take Orders condemned him to earn his living at the Bar between 1846 and 1863, when he was elected into the recently created Sadleirian Chair. It was in this interval that he met J. J. Sylvester and their work on algebraic invariants began. Cayley published over 900 papers on pure and applied mathematics, 27 of them in the Society's journals. He was President from 1869 to 1871.

J. W. L. Glaisher (1848–1928) is perhaps best remembered now by the marvellous ceramic collection he presented to the Fitzwilliam Museum. Thomson called him the most active promoter of pure mathematics in his day, and greatly admired his lectures. He contributed many papers to the Society's journals and filled all its offices in turn, being elected President in 1882.

However, it should not be supposed that either in its election of officers or in its publication of papers the Philosophical Society inclined heavily towards the mathematical and physical sciences. The role of such representatives of the newer, experimental biological sciences as Michael Foster, F. M. Balfour, and Alfred Newton has already been noted. Though in later days one would not find the Society electing as President the Regius Professor of Greek as happened in 1863, or the Lady Margaret Professor of Divinity, as was done in 1867, the older sciences had their fair share of the Society's distinctions. For example, three successive Professors of Anatomy, Sir George Murray Humphry, Alexander Macalister, and J. T. Wilson were all elected Presidents in their turn, though possibly none of them was an absolutely top rank scientist. G. M. Humphry (1820–96) won a knighthood, and has been praised for his influence upon the teaching of medicine in Cambridge; his anatomical papers in the journals seem of a conservative cast, but his books were highly valued in their day and he was a staunch advocate of an

G. D. Liveing 1827–1924   J. Willis Clark 1833–1910

approach to medicine through the Natural Sciences Tripos. Of Macalister (1844–1919), who succeeded Humphry when the latter transferred to a since defunct Chair of Surgery, Gunther wrote: 'He came from Dublin with a wonderful reputation of having navigated the aorta of a whale and of adventurous travels into the interior of seven elephants'. Unfortunately his contributions to the *Proceedings* were far less exciting.

The first experimental science taught in Cambridge was chemistry, but it can hardly be maintained that Cambridge was, at this time, a great centre for chemical research. Only three professors taught in Cambridge between 1815 and 1939. The first, James Cumming (President, 1825–27) was chiefly interested in electricity and the second, George Downing Liveing (1827–1924) in spectroscopy, a subject on which he presented many papers to the Society. He was elected Professor in 1861 and retired in 1908; every Cambridge man must have heard the story of his premature end brought about by collision with a bicycle. Liveing served as Secretary several times and was elected President in 1877. His last paper, read at the meeting of 7 May 1923 when Liveing was 96 (*Proceedings*, 21) dealt with 'The Recuperation of Energy in the Universe', in which he postulated an omnipresent 'universal radiation' to which radi-

ant bodies contributed energy and from which all bodies receive it—a kind of 'steady-state' theory in which there is no
universal increase of entropy. Professor Roughton remembers
this paper as representing the polished language of a bygone age,
but straining somewhat the tact of the proposers of thanks.
Perhaps Liveing's age record among Fellows of the Cambridge
Philosophical Society has been surpassed only by that of the
entomologist Cecil Warburton, whom many will recall as living
in Grantchester to the ripe age of 104.

Also among the leading members of the Society in this turn-
of-the-century period were three zoologists, John Willis Clark
(1833–1910), Adam Sedgwick the younger (1854–1913) and
Arthur Shipley (1861–1927). Willis Clark was for twenty-five
years in charge of the Museum of Zoology, or rather (as Shipley
puts it) 'maid-of-all-work at the Museums, collecting fees for
other people, helping to organise new departments, looking
after his Museum, arranging the Philosophical Library [of
which he was, from 1881, nominally in charge], correcting what
he called the "grosser errors" in the English of the Reports and
other documents sent in by the several Professors or Curators,
settling little disputes between Departments, and generally
"keeping things going." ' From 1891 to his death he was Registrary of the University. His own interest in marine mammals is
reflected in two papers printed by the Society; his address on
the foundation and early history of the Philosophical Society,
given on his retirement as President in 1890, was also printed in
the *Proceedings*, Vol. 7. But he is perhaps best remembered for his
completion and publication of Robert Willis's *Architectural
History of the University and Colleges of Cambridge* (1886). He was a
great benefactor of the Philosophical Library, to which he left
many volumes by his will.

Sir Arthur Shipley (as he became after 1920), author of the
biography of J. W. Clark and his great friend, was a pupil of
F. M. Balfour's as was Adam Sedgwick the younger, who continued Balfour's teaching from 1882 and succeeded Alfred
Newton in the Chair of Zoology in 1907. In the following year
Sedgwick was elected President of the Philosophical Society,
but he resigned both these offices in order to move to the professorship of zoology at the (then recently organised) Imperial
College of Science and Technology, London. He contributed a
number of papers to the Society, mostly embryological. The
Proceedings show, however, that Shipley's work moved away

Sir Arthur Shipley 1861–1927   William Bateson 1861–1926

from embryology to parasitology. A Christ's man will recall
stories of him as a College and University figure, first as Tutor
then as Master; a *bon vivant*, one who loved a Lord and a Prince
even better than a Lord, still for many years a good scientist if
not a great one, who did more than one man's share of teaching,
laborious research, text-book writing and journal editing. He
served the Philosophical Society loyally as Secretary for a
dozen years until he was elected President in 1912; he was a
generous benefactor of his College, and he wrote a number of
readable books apart from his many strictly professional scientific papers.

Lastly, among the exponents of biological science, there were
the two botanists, Francis Darwin (1848–1925) and Harry
Marshal Ward (1854–1906), and not least, the geneticist
William Bateson (1861–1926). Francis, third son of Charles
Darwin and his father's assistant in research during the last
years of the latter's life, contributed greatly to the understanding of Charles's life and work by publishing his autobiography,
correspondence, and the early sketches of the *Origin of Species*.
His own scientific work was devoted to the then novel study of

plant physiology, on which and other botanical topics he lectured in Cambridge from 1884 to 1904. He was also a skilled musician and an essayist. Francis Darwin was elected President of the Society in 1896. Ward had attended Huxley's lectures and spent a year at Owen's College before winning an open science scholarship at Christ's; he worked in Germany and at Kew before spending two years in Ceylon investigating the coffee leaf disease. He was a Fellow of Christ's 1883–85, Professor of Botany at the Royal Indian Engineering College, Coopers Hill 1885–95, and Professor of Botany at Cambridge from 1895. He was active in the organisation of the new Botany School, and President of the Philosophical Society in 1902. His best known work was on trees and their diseases, and on the bacteriology of water; he contributed to the Society his important work on grass rusts.

Bateson is today remembered above all as one of those who took part in the rediscovery and vindication of the work of Gregor Mendel. He is always categorised as a 'biologist', but in fact his early work was in embryology to which he was introduced by Balfour. Travel in the U.S.A. and Egypt aroused in him curiosity about the nature of variation as the key to biological evolution; his book on this subject (1894) already contained the notion of discontinuous evolution, rather than infinitely smooth transition. To the understanding of this Mendel's work, rediscovered (or rather, re-appraised) in 1900 gave him the clue. Bateson devoted himself to the defence of Mendelianism against its critics, his own and his followers' experimental researches playing a large part in the controversy. His presidency of the Philosophical Society came after this battle had been won, and Bateson had been rewarded with a special chair. It was brief, because Bateson left Cambridge in 1910 to bring into existence the John Innes Horticultural Institution, which he made one of the world's foremost biological centres, above all in the science of genetics (Bateson's own term).

Bateson's early work was regularly reported to the Society and briefly noted in *Proceedings*; his main efforts in defence of Mendelianism were directed elsewhere. Two later geneticists both closely associated with the Society were J. B. S. Haldane (1892–1964), Reader in Biochemistry at Cambridge from 1922 to 1932, who contributed nine parts of a monograph on 'A Mathematical Theory of Natural and Artificial Selection' to the *Transactions* and *Proceedings* (1924–32); and Sir Ronald

Fisher (1890–1962), Arthur Balfour Professor of Genetics at Cambridge from 1943 to 1957, who also elaborated the mathematical theory of genetics.

Genetics research at Cambridge was early related to the improvement of agricultural plants, with which Sir Rowland Biffen (1874–1949, Professor of Agricultural Botany, 1908–31) was very closely concerned. Sir Frank Engledow recollects a meeting of the Philosophical Society in the early summer of 1914, in the School of Agriculture and devoted entirely to genetics, at which he (Engledow) presented an account of his own first investigation of a wheat cross. He also worked with George Udny Yule (1871–1951) on a second communication to the same meeting, Yule doing the statistical analysis (*Proceedings*, 7, 1914). Yule had taught applied mathematics for several years at University College, London, before coming to Cambridge as Reader in Statistics in 1912. He was the first to apply statistical analysis to agricultural experiments, and was the Society's President from 1928 to 1930.

Sir Frank also recalls a strange story about Haldane at the Society's dinner in St. John's in 1930 when, demonstrating an efficient way of splitting walnuts without nut-crackers, Haldane employed the table as an anvil and his forehead as a hammer!

When Bateson retired he was succeeded by Sir George Darwin (1845–1912) who had already served as President from 1890 to 1892. He was the second son of the author of the *Origin*; despite unexpected evidences of marked mathematical ability that brought him a Fellowship at Trinity in 1868, he commenced a legal career until deterred by symptoms of ill-health similar to his father's. He started serious scientific work in the late 1870s, and was elected in 1883 to succeed Challis in the Plumian Chair. His research was directed towards the elucidation of the history of the solar system by dynamical reasoning, in particular through study of tidal movements, in the most general sense, and tidal friction. He contributed only one paper to *Proceedings*, on cometary motion.

His son, in turn, Sir Charles Galton Darwin (1887–1962) was Master of Christ's College 1936–38 and Director of the National Physical Laboratory from 1938 to 1949. He had worked with Rutherford at Manchester on the theory of alpha-rays and published an important though little-known theory of X-ray diffraction (1914). At Cambridge in the early nineteen-twenties he worked with R. H. Fowler on atomic statistics and their relation to thermodynamics (*Proceedings*, 21); later he published an influential group of papers dealing with the optical constants of matter and the optical consequences resulting from recent theories of the atom, particularly scattering effects (*Proceedings*, 20, 22; *Transactions*, XXIII). From 1924 to 1936, while he was Tait Professor of Natural Philosophy at Edinburgh, C. G. Darwin did important work in developing the new quantum theory.

CHAPTER FIVE

Meetings, Members and Money

IN THE first days of the Society meetings were held after dinner, in the early evening, since even in the middle part of the last century dinner was often eaten no later than four or five o'clock, and was followed (rather than preceded) by 'tea', as in the novels of Jane Austen. This is apparent, for example, from the account of the meeting in 1867 already quoted. But by 1880 the meeting time (as the hour of dinner advanced) had crept on to 8.30 p.m., provoking Alfred Newton (President) to suggest at the opening meeting of the session that an earlier hour be considered. Accordingly, next month (November) the Society resolved to meet before dinner, at 3 p.m. Probably this time interfered with the 'Grantchester grind' and other forms of afternoon exercise, since in a few years the hour had advanced to 5 p.m. again, 'afternoon tea' being provided at 4.30 p.m. Finally, about 1900, the hour was stabilised at 4.30 p.m. for so long as afternoon meetings of the Society were held.

The number of meetings was somewhat variable. In 1872 the Council agreed that there should be twelve meetings in each year (five in the Michaelmas Term, four in the Lent, and three in the Easter Term), and that there should be a rotation between three 'areas' of subject-interest: (i) mathematics, physics and chemistry (ii) biology and geology (iii) literature, history, law and moral science. (However, it was by this time quite unrealistic to suppose that these non-scientific subjects had any real place within the scope of the Philosophical Society). In later years, it seems, fewer meetings than twelve a year were held in fact, and the Society has certainly never sought to maintain a rigid balance between topics.

In the period after the renunciation of the Society's own House in 1865 meetings took place regularly in a lecture-room

Niels Bohr 1885–1962

adjacent to the Library until after the construction of the Cavendish Laboratory. From that time meetings were occasionally held in the Laboratory. In the 1920s and afterwards meetings were usually organised either in the Cavendish, or in the Old Anatomy Theatre, or in the Botany School.

In the same post-war period the practice of arranging evening lectures in addition to the afternoon meetings was begun. The afternoon papers, sometimes quite brief and nearly always numerous—up to twenty at a meeting, but only a few actually read and discussed—were research reports: regularly the research of young men like P. M. S. Blackett and J. D. Cockcroft was introduced by Lord Rutherford or some other senior Fellow. But the most senior Fellows also presented afternoon communications of their own. The evening speakers came by invitation, and the lecture was a review of a broad field; several speakers were brought from outside Cambridge. Usually there was one of these occasions in each term. Thus in 1921–22

14 November   A. C. Seward: 'Greenland'

22 May   J. Barcroft: 'Physiology of Life in the Andes'.
There was no lecture in the Lent Term because in March Niels Bohr gave a public lecture to the University. In 1922–23 there

were three evening lectures, and the Special Meeting at which Bohr was elected an Honorary Member; in 1923–24 there were again two, and also in 1924–25:

10 November	Sir E. Rutherford: 'The Natural and Artificial Disintegration of Elements'
16 February	D. M. S. Watson (University College, Lonson): 'Orthogenesis'.

This pattern of evening lectures was continued after 1945, but the afternoon communications have not been revived. Thus in 1946 the lectures were

4 March	G. B. B. M. Sutherland: 'Infra-red Detection: Some Wartime Developments and Peace time Applications'
20 May	J. A. Ratcliffe: 'Radar, Drake's Magic Mirror'
28 October	H. Hartridge: 'Bats, and How They Use Sound Waves'.

(Since these things are not recorded, it is not clear unfortunately whether the obvious parallelism between these three post-war lectures was intentional or not.)

In 1947 there was a special lecture (by Alex Wood) on the history of the Cavendish Laboratory, in commemoration of the fiftieth anniversary of J. J. Thomson's discovery of the electron (or more exactly of his determination of the charge/mass ratio of the 'corpuscles' now named electrons), at which the original apparatus was exhibited. It was decided in this year to institute meetings of different kinds, in the various departments, and to bring in their staffs and research students. Two symposia were organised in this year, one on techniques of microscopy and the other on continental drift.* Another innovation of post-second-war years has been the making of visits—to the University Press and University Farm, to a works of the London Brick Company, the University Botanic Garden, Messrs. Chivers at Histon, the plastics factory at Duxford and so on. Perhaps these would have seemed a little frivolous a century ago—though as far back as 1913 a cinematograph lecture was arranged in conjunction with the Marey Institute.

As regards membership, the original rules of 1819 were quite simple: 'That ordinary Members be chosen from the Graduates of this University by ballot.' Only those of M.A. standing could

---

* The subject of an evening lecture in 1923.

be members of the Council; no resident member of the University might attend more than two meetings of the Society without becoming a member of it. Graduates of other Universities might be elected Honorary Members. In November 1872 a new class of membership was created, that of Associates: 'certain persons, not graduates, resident in Cambridge or the immediate neighbourhood.' The Associates were to be elected for a period of three years, which might be renewed, with the privilege of attending meetings and reading in the Library. Clearly the intention was to bring in a number of suitable Cambridge residents, possibly mature men, who could not normally be Fellows. Some availed themselves of this opportunity. But Associate Membership became much more important with the arrival in Cambridge of research students in scientific subjects from other universities, especially those of the Commonwealth and the U.S.A. Such students could not become Fellows until they had graduated (for many years proceeding to the M.A. 'by research', and after 1926 by taking the Ph.D.); in many cases they did not remain in Cambridge long enough to proceed to a degree. Associate Membership brought into the Society many men of the highest scientific distinction—Ernest Rutherford, Niels Bohr, J. Robert Oppenheimer, Hans Bethe and very many others.

From the beginning the pleasant custom was followed of electing distinguished men of science outside Cambridge, English or foreign, to honorary membership of the Society. The first so honoured (in 1819) were Sir Joseph Banks, President of the Royal Society, William Buckland the diluvian geologist, and John Kidd, Professor of Chemistry at Oxford. Sir William Herschel, the great astronomer now advanced in years, was elected in 1820, as was Robert Brown, the botanist, discoverer of the Brownian movement. In electing Edward Troughton the Society paid tribute to the instrument-makers' craft, and he returned the compliment by presenting a barometer and thermometer; other early names are those of the Swedish chemist J. Berzelius, the Scottish 'philosophers' David Brewster and Dugald Stewart, and the French physicist J. B. Biot. Ampère and Faraday were elected in the same year (1823)—it was the latter's first mark of distinction. Cuvier, Gauss, Bessel, von Humboldt, Johannes Müller, Struve, von Helmholtz, Leverrier, J. B. Dumas, Kirchoff and many other distinguished foreigners had been honoured by mid-century. Among more recent

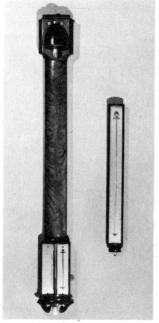

Barometer and Thermometer given in 1820 by Edward Troughton, one of the first Honorary Members.

H. L. F. von Helmholtz 1821–1894

familiar names one sees those of Hertz, Mendeleef, Michelson, Poincaré, de Vries, Max Planck, Maurice de Broglie and Niels Bohr. The election of the last-named as an Honorary Fellow on 12 June 1923 (he had been admitted as an Associate in 1911) was made the occasion of a Special Meeting in the Cavendish Laboratory. Bohr himself was expected to be present to receive the honour, and it may appear from the following document (presumably spoken by the President, C. T. Heycock) that this was indeed the case:

Professor Niels Bohr. Professor of Theoretical Physics in the University of Copenhagen.

His researches are well known to applied mathematicians, physicists and chemists. To mathematicians amongst other things for the detailed applications of the dynamical theory of perturbations which he has made to atomic problems. To the physicists for his beautiful applications of Sir E. Rutherford's nuclear atom and Planche's [sic] quantum theory of radiation to explain first the spectrum of hydrogen and then the general

Letter from Michael Faraday, 1823, acknowledging Honorary Membership.

features of all atomic, X-ray and optical spectra. To chemists and physicists alike for his brilliant work in relating in the most fundamental and convincing way the periodic table of the elements with the electronic structure of the corresponding atoms.

It is only just over ten years since Prof. Bohr first put forward his theory of the hydrogen spectrum. It is interesting to contemplate the revolutionary changes which have arisen as a result of this theory in our views of atomic physics.

Spectroscopy was a moribund side line; it is now the very heart and centre of physics and a new spectrum is disentangled every month. We have a wholly new fundamental branch of physics in the study of electronic impacts. We can visualise a concrete model of the atom with confidence that it represents at least the essential features of the truth. The domain of application of mechanics has been vastly extended and at the same time the non-mechanical nature of physical and chemical processes becomes clear.

We welcome Professor Bohr.

(The name in line eight should of course be that of Max Planck, another Honorary Member). Bohr received in the following year a further honour from the Philosophical Society in the publication as a supplement to the *Proceedings* of an English translation of his monograph *On the Application of the Quantum Theory to Atomic Structure. Part One*, which had previously appeared in the *Zeitschrift für Physik*.

Since the last election of Honorary Members took place some forty years ago it cannot be a matter for wonder, though it is to be regretted, that none are alive at the time of writing this account.

When a large group of Honorary Members was elected in 1914 the opportunity was taken to include Madame Curie's name in the list, after it had been ascertained that this was a proper step, since at that time women could not become Fellows. Legal opinion held that as Honorary Members had no rights, the Bye-laws were not infringed. The position of women was indeed anomalous, for they could also be admitted as Associates, and had been for some years as a matter of routine. Indeed, one could hardly argue that the Society was hostile to women. They had been admitted to meetings in the 1860s, and probably much earlier. On 26th February 1883 Miss Alice Johnson of Newnham had communicated a paper 'On the development of the pelvic girdle and skeleton of the hind limb in the chick', later printed in the *Proceedings*, seemingly in her own person since there is no record to the contrary. However, when other women communicated papers in the pre-1914 years it seems that this was done by a Fellow, or a male collaborator, or else the paper was 'taken as read'.

Through all these years women were barred from Fellowship by the simple fact of their lack of degrees, whatever their eminence. Finally, in November 1928, Dr. Joseph Needham

proposed that the Bye-laws should be amended in order to permit Titular Graduates to be Fellows and Council members; this reform with some others was effected at a special General Meeting of the Society on 4 January 1929. It may be added that the first, and only, Ladies' Dinner organised by the Society was held in 1938.

The number of Fellows has passed through quite marked variations (graph p. 71). Surprisingly, the Society rapidly attained a membership of two hundred and through the middle years of the nineteenth century passed the five hundred mark, a figure not to be exceeded before 1930. From about 1870 until the early years of this century the membership declined, without any obvious explanation. Perhaps the sale of the house in All Saints' Passage made the Philosophical Society much less attractive as a club, and after 1881 the Library was open to all for reference; the virtually exclusive concentration of its interest on science would certainly tend to exclude the non-scientific graduates of the University. Yet, at this same time, science was flourishing in Cambridge as never before (especially after 1880) and there were far more scientific men in Cambridge than ever in the past. It may be remarked too that the Council minutes for this period are singularly dull and lacking in incident, which may indicate a lack of vigour in the Society itself, despite the number of eminent scientists enrolled in it. It would be interesting to be able to compare the sizes of its audiences at different periods.

In the First War decade 1910–1920 the Society's Fellowship again declined slightly, a phenomenon not repeated during the Second World War, though (for unknown reasons, other than enforced departure from Cambridge) there was an unusually large number of resignations of both Fellows and Associates at the beginning of the 1939–40 session. After the war there was a considerable recruiting drive—in 1946 a direct approach was made to all appropriate University teachers who were not Fellows already asking if they would like to join the Society— with such success that by 1948 the 1939 total was exceeded. In 1968 the number of Fellows was 807.

Only in quite recent years has the Philosophical Society ceased to be impoverished. In the nineteenth century there was no real recovery from the situation that forced the sale of its house in 1865. Publication, even with generous assistance from the University Press, absorbed all the Society's revenue, and no

fresh ventures were possible. In 1888, in which year the accounts showed a surplus of some £40, the Society asked the Press Syndicate to assume the whole cost of printing its *Proceedings*, a request the Syndics quite reasonably declined. They were already printing the *Transactions* without charge. However, the Museums and Lecture Rooms Syndicate came to the assistance of the Society by accepting responsibility for the purchase of certain periodicals.

A guinea subscription gave good value to Fellows and the Society could limp along, with a few hundred in reserve, so long as the guinea retained its purchasing power, despite decreasing numbers. Through, and after, the 1914–18 war the situation became impossible. By 1922, only three years after a grand dinner celebrating the completion of the Philosophical Society's first hundred years, it was in serious financial difficulties. The sums involved were not large—a deficit on the year's working of £123—but were larger than appeals to the membership could easily remedy. Composition fees had for many years been expended as current income; costs had gone up (publication costs rose from £167 in 1913 to £420 in 1920). This expense of publication was of course essential because most of the 600 journals received by the Society were obtained by exchange.

A Memorandum by the President (A. C. Seward) rightly insists on the importance of maintaining the exchange system for the benefit of the Society's Library:

At present the Philosophical Library receives about 600 journals (representing about 2100 parts of volumes). All of these, with the exception of 30 which are given and 15 which are purchased, are sent in exchange for our publications. Of these about half (300) are not taken by the University Library. It is impossible without great labour to estimate the value of the yearly accessions to the Library, but even if we accept the low figure of 5/- per part as an average price the annual value so given of £500 far outweighs the cost of the periodicals sent in exchange. Apart from their value as a medium for publishing the scientific researches of members of the University, the publications of the Society are thus a distinct financial asset to the University, in relieving the University Library from the necessity of subscribing to a large number of foreign periodicals.

It is estimated that an average of 50 persons use the Library each day in term.

The total debt was less than £300, though it would have been greater if binding had not been postponed and so on. As remedial measures the subscription was doubled to two guineas and other measures of economy adopted, including, regrettably, the sale of books. One victim, sold in 1924 after much heart-searching, was James Audubon's *Birds of America*, 1827–38, to which the Society had been one of the original subscribers. It fetched £400, a fraction of its modern value.

In the bleakest period of the Society's history the University Press wrote off a deficit on the printing account and provided a subsidy of £125 a year; the University contributed generously to the cost of binding periodicals; and the Royal Society made a grant of £150 per annum in 1925 and 1927, and again from 1927 onwards, in aid of the Society's journals. In 1928 the Royal Society at first refused to make a grant, much to the Society's indignation; it later appeared that the Royal Society felt that it could not in any way aid the publication of the Philosophical Society's *Transactions*, whether because of their miscellaneous character or because the University Press had long printed them without charge. As a result, the *Transactions* of that year were the last to be issued, though a decision finally to abandon their publication was postponed to May 1931.

Meanwhile, the Royal Society had relented and paid the grant for 1928, repeated annually until 1931; in 1937 there was a grant of £100, and larger sums were received in the first decade following the War. The University Press, in lieu of printing the *Transactions* gratis, made an annual grant of £125, also enlarged after the War until it was replaced by a 'rebate allowance' in 1961. In recent years (since 1957) the Philosophical Society has further been aided by a substantial annual grant from the University in recognition of its contribution to the Scientific Periodicals Library.

As a result of such generous assistance and the prudent measures adopted in the late nineteen-twenties the Society's financial position improved enormously. Early in 1930 the total deficit incurred by the Society had been reduced to £253; the year 1932 showed the handsome excess of income over expenditure of £464, and the Society's invested capital amounted to £3,355. Successful publication was the key to this success, as may be seen from the figures produced by the Treasurer, F. A. Potts, in 1931:

		£
Receipts from Fellows receiving *Proceedings*		480
Other receipts from *Proceedings*		490
		970
Less printing costs		677
'Profit'		293
Receipts from Fellows receiving *Biological Reviews*	280	
Other receipts from *Biological Reviews*	487	
		767
Less printing costs		589
'Profit'		178

(The Royal Society grant of £150 was not taken into account.)*

From this time the finances of the Society have continually increased in strength, enabling it to expand its activities. By 1937, with a total income approaching £2,500 per annum, it was possible to show a favourable balance on the year's working of £650 after meeting a printing bill of £1,700 and other expenses. Since the war general inflation, the success of the Society's journals, and a wise policy of investment have combined to multiply these figures many times:

Year 1954

Income		Expenditure	
Subscriptions	£1,240	Editorial and printing	£4,776
Sales	6,104	Other	301
Grant from Press	200	Surplus	3,068
Royal Society grant	150		
Investments	451		
	£8,145		£8,145

At this point there was one obvious weakness: the funds of the Society, valued at £11,403, actually stood at a *lower* book-value than in 1938 despite the post war inflation, and the average yield was under 4%. Accordingly in that year the Society decided to transfer part of its capital to equities and to employ professional advice in managing its portfolio. The effect of this change, together with continued investment of surpluses, was to multiply the Society's capital ten times, and to increase its income from investments eleven times. Meanwhile over the same decade, though the income from subscriptions grew little,

---

* The average costs of the journals over the last three years were somewhat different, £818 + 479 = £1,297—only a little discrepant from the total above.

that from sales of the journals almost doubled. By 1965 the income from these two sources still only just covered the cost of editing, printing and distributing the journals, but the Society's considerable income from other sources frees it from fear of penury.

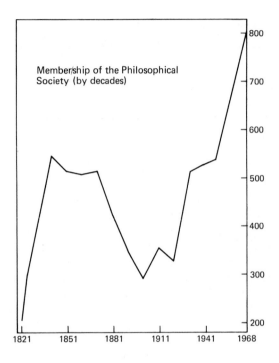

Membership of the Philosophical
Society (by decades)

CHAPTER SIX

New Physics
and
New Mathematics

SIR GEORGE Darwin was one of the last mathematical physicists of the classical school, fully in the Cambridge tradition; his son was a member of the first generation to adopt the new physics, especially the quantum mechanics of Bohr and Heisenberg. As is well known, Niels Bohr was closely associated with the 'Cambridge School'—and even more with Ernest Rutherford. The latter did not, of course, work in Cambridge between 1898 and 1919, but he was always a member of the 'Cambridge School', corresponding regularly with J. J. Thomson (1856–1940) who, with his later associates including F. W. Aston, continued to study the atom experimentally and to devise mathematical models of its structure.

Modern physics began in Cambridge with J. J.—in England as a whole it might be said to have begun with Sir William Crookes, who was invited to lecture on his apparatus by the Philosophical Society in May 1879. But something should first be said of two men who remained closer to the classical tradition. A. E. H. Love (1863–1940) left Cambridge in 1898 to become Sedleian Professor of Natural Philosophy at Oxford, after presenting to the Society several papers on his special interest, the theory of elasticity. Sir Joseph Larmor (1857–1942), however, only briefly left Cambridge (1880–1885). He was one of the Society's Secretaries from 1886 to 1895, and President from 1898 to 1900; in 1903 he succeeded Stokes in the Lucasian Chair. Joseph Larmor had been Senior Wrangler in the year that J. J. Thomson was second; their life-work in the experimental and theoretical parts of physics was to some extent parallel. Larmor's first great endeavour was to describe (see *Aether and Matter*, 1900) a development of electromagnetic theory by which matter was composed of 'electrons' which were themselves 'freely mobile singular points in the specification of

Sir Joseph Larmor 1857–1942

the aethereal strain.' Among later achievements he produced a theoretical account of the Zeeman effect, examined the problem of the reflection of radio waves in the upper atmosphere (*see also*, p. 74), and attacked the problem of aether-drift (the negative result of the Michelson-Morley experiment) on lines similar to those of Lorentz. Larmor, it has been said, stood between the old and the new physics; on the one hand he was one of the last great exponents of classical mechanics, while on the other certain aspects of his work tended towards the establishment of relativistic ideas; moreover, he did himself (as late as 1937) publish in *Procceedings* a paper 'On temperature in relation to quantal phenomena.' Always he applied a critical mind to twentieth-century innovations. His fundamentally conservative cast of mind in later life was well displayed in his attitude towards a proposal to introduce a bath-house to St.

Sir J. J. Thomson 1856–1940

John's College in 1920: 'We have done without them for 400 years; why begin now?' His adaptability is no less shown by his regular use of the College bath-house in the latter part of his bachelor existence in College.

Larmor impressed younger men, like J. A. Ratcliffe and F. J. W. Roughton, by the fertility of his scientific thought. The former vividly remembers Larmor's communication on the bending of radio waves, actually evolved while he was giving lectures on the subject (27 October 1924), the latter writes of a meeting in the same year at which research by Hamilton Hartridge and himself was presented on the measurement of the velocity of very rapid chemical reactions (*Proceedings*, 22, 1924 and 23, 1926), 'which proved a curtain-raiser to much new and interesting work.' Larmor, Professor Roughton continues, 'who was so interested in the Society and attended most of its meetings, was present at that particular meeting and sent us next day four closely-written pages of comments, of which unfortunately we could barely read a word. Larmor's handwriting was notoriously illegible.'

As Larmor's lifetime spanned the transition from classical mechanics to the concepts of quanta and relativity, so J. J. Thomson's traversed the modern history of experimental physics. When he was an undergraduate there was virtually no experimental training, apart from Stokes's optical demonstrations. At the Cavendish too experimental work was, at first, of a fairly routine character—determination of constants, some spectroscopy, construction of galvanometers and electrometers, pendulum experiments and so on, though as early as 1874 W. M. Hicks, later Professor at Sheffield, attempted a demonstration of the existence of Maxwell's electromagnetic waves. The real change came when J.J. began, about 1885, to study with Richard Threlfall the passage of electricity through gases, that is, to work for the first time in Cambridge with discharge-tubes (see *Proceedings*, 1886, 1889, 1891 etc.). Previously J.J.'s published work had been nearly all mathematical, though he had done some work at the Cavendish on electromagnetic induction, starting in 1880, the year of his Tripos. The vacuum-pump (improvements in which contributed so largely to J.J.'s own successful determinations) has indeed been the primary tool of modern physics, since it was essential to the development of experiments on particles and to the evolution of the thermionic tube.

In the last years of Sir Joseph Thomson's lifetime, when he was Master of Trinity, the most important and enduring subatomic particles had already been discovered, accelerators of high energy had been developed, and the possibility of obtaining energy from nuclear fission was known.

To the dissolution of the 'billiard-ball' atom—and it is easy to exaggerate both the inviolability and the credibility of such an atom in the minds of late-nineteenth century scientists—J.J. made a most important contribution by his measurement of the e/m ratio of particles in the 'cathode rays'. This not only proved decisively that the 'rays' were composed of particles rather than waves, but suggested that the particles possessed about one-thousandth of the mass of the hydrogen atom; that is, they were particles of sub-atomic size. J. J. first described these experiments at the Royal Institution on 29 April 1897 (see also *Proceedings*, 9, 1897). They did not, indeed could not, suggest immediate new ideas about the structure of the atom, especially as it was customary at this time to think of the passage of electricity through gases and liquids as strictly analogous. But the 'dis-

Lord Rutherford 1871–1937

covery of the electron' was the most positive piece of information in the wealth of new knowledge about the electrical properties of matter emerging at this time. Röntgen's discovery of
X-rays in 1895 and Becquerel's of radioactivity in 1896 had
already confounded established views of matter, radiation, and
energy. (Some early Cambridge reaction to these new phenomena may be found in J.J's first paper in *Proceedings* on X-rays
[9, 1896] and Stokes's 'On the nature of the Röntgen rays' [9,
1897]). As everywhere the mysterious powers of the X-rays
were demonstrated by photographing the bones of the living
hand, coins in a box and so on; it was J.J. who discovered with
delight that X-rays ionised gases.

But the solid work was on the electrical conductivity of gases, in which Townsend, Rutherford and many others joined. Between 1896 and 1900 a total of over a hundred papers came from the Cavendish Laboratory. These last years of the nineteenth century were those in which Cambridge recognised the concept of 'research student' at last, and in physics they began to come to the Cavendish in numbers. Few of them realised the significance of Becquerel's discovery; indeed, its discoverer soon renounced it as insoluble. The first references to uranium in the Cambridge Philosophical Society occur in two communications on the relation of uranium radiation to the formation of clouds, by C. T. R. Wilson of 25 October and 22 November 1897; his deep love was for meteorology, and therefore he was the less involved with experiments on the conductivity of gases and others involving the discharge-tube. (C.T.R. made numerous communications to the Society in the early years of the new century, but few in subsequent years when he was President, 1918–1920, or later still when he occupied the Jacksonian Chair.) Similarly the most famous of all these early research students, Rutherford, worked first of all on magnetic hysteresis, and the electromagnetic-wave detector he derived from it. It was only just before leaving Cambridge in 1898 and at Montreal that Ernest Rutherford (1871–1937) turned seriously to radioactivity at a time when almost everyone else, save the Curies, had lost interest in it.

Rutherford's work on thorium, the discovery and identification of alpha and beta radiations, the bold proclamation of atomic transmutation and the first atomic model have nothing to do with this story. Rutherford published only one early paper in the *Proceedings*, on ultraviolet light as a cause of ionisation (9, 1898). After his departure the study of ionisation in general, and in particular of the conduction of electricity through ionised gases, remained the main line of work at the Cavendish, and was the subject of J.J.'S great book of 1903. But from 1902 onwards radioactivity and atomic physics came increasingly to the fore, no doubt as a result of the publication of the theory of Rutherford and Soddy. During the first decade of the twentieth century there were 137 people in all working at the Cavendish for longer or shorter periods, that is, about 25 or 30 in any one year. The total number of papers printed on an average in each year was rather more, say 30–35. In 1904 only four papers, all by J. J. Thomson, deal with atomic physics. In later years there might

77

be as many as eight or ten on this subject, with more widely diversified authorship, printed in various journals. The connection with the Philosophical Society was very strong, since many of the Cavendish workers were Fellows, and about a third of all their papers was published in its *Proceedings*—53 out of the total of 162 papers coming from the Cavendish in the five years 1900–1904 inclusive.

In 1904 Charles G. Barkla (later Professor at King's College, London) measured the scattering of X-rays through several gases and, using a theory due to Larmor, showed that the number of electrons in each atom was approximately equal to half the atomic weight (except in the case of hydrogen, with one electron); and in the same year J.J. described in the *Philosophical Magazine* his theoretical model of a 'plum-pudding' atom. He contributed a supplementary paper on this to *Proceedings*, 13, in 1905. As is well known, his own later work was mostly devoted to the examination of the 'positive rays' (Kanalstrahlen) by his parabola method, which led on to the mass spectrograph of Aston and his work on isotopes (to which a couple of papers by him in *Proceedings* relate).

J. J. Thomson himself published twenty-five papers in the same journal during the ten years 1904–13 inclusive, on such topics as 'The electric theory of gravitation', 'The nature of the gamma-rays', and 'The unit theory of light' as well as on X, cathode and positive rays. C. T. R. Wilson in the same period provided three communications dealing with radiation and others came from N. R. Campbell, a Fellow of Trinity who later went into industry, J. A. Crowther of St. John's, Frank Horton (for long Professor in London University), R. D. Kleeman and John Satterly (respectively research students from Australia and the Royal College of Science), and Richard Whiddington who was to move in 1919 to the Cavendish Professorship of Physics at Leeds. All of these were, of course, research workers under J.J.

After the First War, when Rutherford had returned to the Cavendish as J.J.'s successor, a fresh start was made, carrying on from Rutherford's enormously successful pre-war period in Manchester, and inaugurating a great phase in the history of the Cavendish Laboratory which is, in turn, reflected in that of the Philosophical Society. Not that Rutherford sent it much for publication—one paper of his own in 1923, one with W. A. Wooster in 1925, and a third with James Chadwick (who was

Rutherford's major collaborator in the 1920s) on 'Energy relations in artificial disintegrations' (1929). But he attracted brilliant young men to Cambridge (as he had previously to Manchester), he encouraged them to become Associates and Fellows of the Society, and very often himself communicated their first important results to its meetings.

By the early 1920s atomic physics was dividing into two streams, that of the experimenters and that of the theoreticians. The latter were represented (so far as Cambridge and the Philosophical Society are concerned) particularly by Niels Bohr (1885–1962), C. G. Darwin (who had worked with Rutherford and Moseley at Manchester) and R. H. Fowler (1889–1944), together with the younger men P. A. M. Dirac, D. R. Hartree (1897–1958), N. F. Mott, C. A. Coulson and R. E. Peierls, all of whom figure in the *Proceedings*, some extensively. In the first decade of the twentieth century atomic models (including Rutherford's of 1911) imposed no extraordinary demands on mathematical abilities; the change came with the introduction of quantum concepts by which the electron composition of an atom was enabled to account for its chemical and spectroscopic properties. Bohr's great initial work in this direction (1913) attracted little serious attention. He had arrived in Cambridge with his Ph.D. dissertation on the electron theory of metals in 1911, without winning much response from J. J. Thomson; meeting Rutherford there (whom he was to describe as 'almost a second father to me') Bohr had gone to join him at Manchester, where he proved a most useful member of Rutherford's group, which Bohr often rejoined after his return to Copenhagen in July 1912. In the following year his quantum theory of the hydrogen atom was discussed with Rutherford, who considered it too elaborate. But for the war Niels Bohr might have settled in England; however, when an invitation to take up a Royal Society Professorship at Cambridge was extended to him (1923) it was already too late. Nearly all who worked on the quantum theory of atomic structure at Cambridge in the twenties and thirties visited Bohr's Institute at Copenhagen.

By the early 1920s this, the 'old' quantum theory, had attracted considerable interest because of the growing correspondence between theory and experimental results, but serious problems in its extension were encountered which led to the introduction of the matrix mechanics of Heisenberg (1925) and the wave-mechanics of Schrödinger (1926). Most of this stage of the development of theoretical atomic physics belonged to continental Europe.

Of the Cambridge group, R. H. Fowler and D. R. Hartree were first in the field with major articles in *Proceedings* (1923) respectively on the theory of the motion of alpha-particles through matter, and on numerical applications of Bohr's theory of spectra. Both authors contributed many more papers on atomic theory,* and the latter's theory of the self-consistent field was very important. But the great mind at work on quantum theory, in the *Proceedings* as elsewhere, was that of Dirac, whose epoch-making paper on 'The quantum theory of the electron' appeared in 1928; this contained the first hypothetical suggestion (in the existence of a positive as well as a negative solution to his equations) of a particle other than the already familiar proton-electron pair; the actual existence of the positron was first revealed, independently, by Carl D. Anderson in August 1932. Dirac's book *The Principles of Quantum Mechanics* appeared in 1930, and he returned to the positron in another *Proceedings* paper of 1934.

On the experimental side the Cavendish school headed by Rutherford was outstanding. In the same year, 1932, James Chadwick amassed the direct evidence for the existence of a fourth fundamental particle, the neutron (which, however, he at first proposed as a proton-electron pair). Throughout these years the work was mainly on the effect of alpha-particles in disintegrating atomic nuclei, and later on experiments with neutrons for the same purpose. A great many of the results were published in the Philosophical Society's *Proceedings*, for example, two neutron papers with Chadwick as co-author in 1934 and 1935. In the years 1923 to 1927 inclusive 47 Cavendish authors (or collaborations, the number of individuals involved is fewer) published a total of 88 papers, of which 27 appeared in the Philosophical Society's *Proceedings*. Many of these, and summarised communications, came from the younger workers at the Cavendish. At this time, as earlier, the other periodicals to which papers were commonly sent were *The Proceedings of the Royal Society* and the *Philosophical Magazine*. A rough count of references in the book *Radiations from Radioactive Substances* (1930)

---

* It was only in 1934 that Hartree began to interest himself in computing techniques, his main concern after his return to Cambridge as Plummer Professor in 1946.

C. T. R. Wilson 1869–1959

by Rutherford, Chadwick and C. D. Ellis suggests that about 48% of the references were to *Proc. R.S.*, 35% to the *Phil. Mag.*, and 17% to the *Proc. C.P.S.*, but in the last two sections of the book, reflecting the recent work at the Cavendish, the references to each of the three periodicals are about equal in number.

To turn over the list of contributors on topics in nuclear physics to the *Proceedings* during the 1920s and 1930s is, in fact, to recite the names of many of those who have occupied a leading place in British physics during the last decades: P. M. S. Blackett, E. C. Bullard, J. Chadwick, J. A. Chalmers, P. I. Dee, C. D. Ellis, K. G. Emeléus, N. Feather, H. Jones, Sir Harrie Massey, P. B. Moon, M. L. Oliphant, Sir George Thomson, and E. J. Williams, for example, besides Rutherford himself. A few words on some others may perhaps be permitted.

First, Charles Thomson Rees Wilson (1869–1959), who was an experimental scientist of the highest distinction, tracing his career back to two exceptional and stimulating meteorological experiences in Scotland of 1894 and 1895. These induced him to design his 'cloud chamber' in order to study the formation of droplets; in 1897 he showed that ions produced by irradiation could serve as nuclei for condensation. However, perhaps because of his own more immediate concern with atmospheric electricity, it was not until 1911 that the first photographs of particle-tracks were taken. It was a technique that Wilson advanced to great perfection. Professor Blackett recalls that in 1921 Rutherford set him to photographing, in the cloud chamber, the disintegration of atomic nuclei by alpha-particles (this was achieved in 1924); 'So of all my generation', he writes, 'I perhaps am the most deeply indebted to C. T. R.'s genius which enabled him to perfect the first method of revealing the tracks of individual particles.'

C. T. R. Wilson came to work under J. J. Thomson in 1895; he taught in Cambridge from 1900 to 1934, having been Jacksonian Professor for the last ten years. Many still recall how enthusiasm for research was generated in his Part II laboratory class. Much of his work was described to the Philosophical Society, of which he was President from 1918 to 1920, but it was only briefly noted in the *Proceedings* because of publication elsewhere—for example, his research of 1901 on the spontaneous ionisation of gases which led to the discovery of cosmic rays (*Proceedings* 11, 1901; *Proc. R.S.* LXIX).

Secondly, a great teacher at the Cavendish, George Frederick Searle (1864–1954), whose Part I practical class is still vividly recalled by many physicists of the pre-1939 generation, had an even longer career of over fifty-five years. Beginning his work at the Cavendish in J.J.'s early days (1888), he was brought back from retirement to assist during the 1939–45 war, with his gift for devising brilliant teaching experiments unimpared. He contributed many papers to *Proceedings* between 1892 and 1936 on a great variety of topics—from experiments on the surface tension of soap films through a determination of the focal length of a thick mirror to the calculation of resistance-networks. In fact, these were his teaching experiments and undergraduates worked from the printed texts.

The Society received many papers from the Cavendish at this time, it met there frequently, and its officers regularly came from the Cavendish group. F. W. Aston, as Secretary from 1920 to 1925, and R. H. Fowler (Secretary, 1925–28) were active in bringing good work before the Society, while J. D. Cockcroft, Secretary from 1939 to 1926 and Treasurer from 1937 to 1945, was another loyal supporter of the Society, though he himself

F. W. Aston 1877–1945   Sir John Cockcroft 1897–1968

submitted for publication only an early paper on transformer core heating (*Proceedings* 22, 1925). There is, however, a reminder of the celebrated proton-accelerator of 1929 in the paper by Cockcroft's collaborator, E. T. S. Walton, on 'The production of high speed electrons by indirect means' (*Proceedings* 25, 1929).

Atomic physics was not the only common ground between the Cavendish and the Philosophical Society. In the early days of J. J. Thomson, Ambrose Fleming had been a student at the Cavendish; he had since joined Marconi, and his interest in the emission of electrons from hot filaments, as we should now say, brought him to the invention of the thermionic diode in 1904, which Lee De Forest improved into the triode in 1909. By 1920 the circuitry employed in radio was quite advanced and the power radiated high; voice telephony had crossed the Atlantic. But neither the action of the thermionic valve nor the propagation of radio waves was well understood. E. V. Appleton (1928–1965), who had begun work on crystallography under Sir William Bragg, determined to tackle both these problems as a result of his wartime experience. Generously, Rutherford agreed to support this research at the Cavendish; it produced confirmation (1924) of the existence of the predicted 'Heaviside' reflecting layer in the ionosphere, and proof (1929) of the existence of a higher, 'Appleton' layer. In 1936, after a dozen years' absence, Appleton returned to Cambridge as successor to

H. F. Newall 1857–1944

C. T. R. Wilson, but was soon called away to the DSIR, where he did invaluable organising work during the War. A good deal of Appleton's work, and that of his associates and successors in research into ionospheric physics (among them M. A. F. Barnett, J. A. Ratcliffe and K. G. Budden) can be traced in the Philosophical Society's *Proceedings*.

Radio merges into radar (but the great Cavendish war-time effort on that, led by Cockcroft, was no concern of the Philosophical Society) and this, as is well known, into radio astronomy, in which branch of science Cambridge has become a centre of great eminence. There is not space more than to hint at the achievements in optical astronomy and stellar spectroscopy of such men as H. F. Newall (President, 1914–16), who had intended to pursue physics under J.J. until his father's death left him with a 25-inch telescope on his hands, and his successor in the Chair of Astrophysics, F. J. M. Stratton (President, 1930–31). Long before this time it had become clear that Cambridge was not one of the world's most fortunate sites for optical instruments. However, for theoretical astronomers it offers great intellectual advantages. The Newtonian tradition revived by J. C. Adams and G. H. Darwin was strikingly extended to the stars by J. H. Jeans and by Sir Arthur Eddington (1882–1944), who contributed four widely scattered papers to

Proceedings. As Plumian Professor (1913) Eddington also took over direction of the Observatories. 'The modern Archimedes'— as Whittaker termed him—occupied himself with the general theory of relativity from 1916, and in the 1920s became equally aware of the broadest significance of the quantum theory. His posthumous book, *The Fundamental Theory* (1946), was the fruit of his long endeavour to synthesize, in a single derivation, both these great branches of modern theoretical physics.

'One of the greatest figures in the greatest period of development of modern astrophysics and cosmology', E. A. Milne (1896–1950) was a generation junior to Eddington, partly because his career in pure science—he was also extremely able in applied—was delayed by the War. Milne sent several papers to *Proceedings* in the 1920s, although he moved to Manchester in 1924, and to Oxford in 1928. This was the time when Eddington, Milne, Fowler and others were bringing yet another branch of physical science in Britain to a position of international eminence, upon the foundations laid long before by the spectroscopists, Einstein, and De Sitter. To this the new concepts of radiation and the mathematical theory of the Rutherford-Bohr atom, together with the experimental discoveries of the particle physicists, were all alike vitally necessary. And with the work of H. Bondi, T. Gold, F. Hoyle, and R. A. Lyttleton on the theoretical side, with the radio astronomers on the experimental, investigation of the universe on the large scale also continues to be reflected in the *Proceedings*.

Much else might be said here; of the far-reaching appeal of the Cavendish, for example, which again one finds manifest in the *Proceedings* through the names of Hans Bethe, Max Born, C. J. Eliezer, P. L. Kapitza and J. R. Oppenheimer. Here the four quarters of the world were called together by Rutherford's genius in peace, some to reassemble only once again at Los Alamos in war. Nor can justice be done to the important lines of investigation requiring classical methods of investigation into problems of applied mathematics, carried on with great distinction through this period into the post-war era; the names of Sir Harold Jeffreys and Sir Geoffrey Ingram Taylor (President, 1967–8), authors of many contributions to *Proceedings*, may here be representative of others also.

The pure mathematicians in the Philosophical Society took the leading part in the revitalization of their subject in England that occurred early in the twentieth century. Despite various nineteenth-century reforms and the particular eminence in his own line of Arthur Cayley the study and teaching of mathematics in Cambridge had assumed an outmoded and narrow character by the eighteen-nineties. The new ideas and new branches of mathematics thriving on the continent, in France especially, were hardly at all followed in England, and therefore not in Cambridge, still the leading centre for mathematical studies. This was in part an unfortunate consequence of the stringent conditions and highly competitive spirit of the Mathematical Tripos, an examination which had always favoured speed and facility rather than depth of understanding and creative ability. And it was distinction in the Tripos that won men election into Fellowships, and the possibility of taking up mathematics as a career. J. J. Thomson in his autobiography provided a sharp picture of the Mathematical Tripos as it was in the old days—he took it in 1880—three days of 'book-work' in which use of the calculus and analytical geometry was forbidden; another day in which these methods were allowed (and questions on physics were included); then, after an interval of ten days, five more days in which the candidates deemed worthy of honours were examined again. These were the days of competition, when ability to write down the 'book-work' speedily and accurately brought high marks: 'Accuracy in manipulation was perhaps the most important condition in this part of the examination, and the most difficult to impart' wrote J.J., who regarded it as an excellent preparation for the Bar! But J.J., while seeing its weaknesses, nevertheless regarded the old Tripos as a good test of mathematical quality. Curiously, like so many first-rate scientists (including Maxwell) he had been *second* Wrangler.

The way in which the old Tripos selected men for posts outside Cambridge, as well as within, can be illustrated by referring briefly to the lives of four men, each of whom made important contributions to the Philosophical Society's publications, as well as in other ways: William Burnside, Professor of Mathematics at the Royal Naval College to 1919; M. J. Hill, who held London Chairs from 1884 to 1923; George Ballard Mathews, Senior Wrangler of 1883, who taught at Bangor for some years and later returned to Cambridge; and A. C. Dixon, Senior Wrangler in 1886, Professor in Northern Ireland colleges from 1893–1930. The Senior Wrangler was not invariably a great mathematician, but it was virtually certain that he could, if he wished, be an influential one: through his work on the theory of groups Burnside especially had a strong significance for current work in mathematics. (To the best mathematician of 1890, however, Hobson's pupil Miss Phillippa Fawcett, the possibility was less equally open).

1909 was the last year of the old Tripos; his friends hoped that C. G. Darwin would be the ultimate Senior Wrangler, but he was in fact placed Fourth. Of him Sir George Thomson has written:

> The teaching of mathematics in those days in Cambridge was decidedly conventional, and though he [Darwin] records his debt to the 'invaluable drill' given by the coaching of R. A. Herman, he criticised severely in after life the deficiencies of the syllabus which was disconnected from the subjects then coming into importance. Late in life he wrote that he had never heard of relativity or the quantum theory before he left in 1910.

That relates to the applied side. The modernisation of pure mathematics in Cambridge had begun many years before, an early landmark being the publication of A. R. Forsyth's *Theory of Functions of a Complex Variable* in 1893. Although Sir Edmund Whittaker wrote of this book that it 'had a greater influence on British mathematics than any work since Newton's *Principia*', and although it introduced continental ideas new to most British mathematicians, Forsyth's treatment was so lacking in rigour that it can hardly be regarded as laying secure foundations for the development of modern pure mathematics in Cambridge. It was certainly no less significant that a gifted young mathematician like G. H. Hardy, while still an undergraduate, should in this same period be encouraged by A. E. H. Love to read the *Cours d'analyse de l'Ecole Polytechnique* (1882–1887) of Camille Jordan (1838–1922) where he found a modern treatment of the theory of functions; 'I shall never forget', he wrote, 'the astonishment with which I read that remarkable work, the first inspiration for so many mathematicians of my generation, and learnt for the first time what mathematics really meant.'

Andrew Russell Forsyth (1858–1942), Secretary in 1889, taught at Trinity from 1884 to 1895 when he was elected into the Sadleirian Professorship, which he resigned in 1910 for personal reasons. For ten years after 1913 he directed the Mathematics Department at Imperial College, London. Forsyth was active

E. W. Hobson 1856–1933   H. F. Baker 1866–1956

on the Council of the Society during many years, and contributed a number of papers to the *Transactions* on the theory of differential equations (1889–1900), on which he published a four-volume book. He also took part in the preparation of Cayley's collected works, and edited the *Quarterly Journal of Mathematics* from 1884 to 1895.

Higher standards of rigour and a no less creative familiarity with modern mathematical work on the continent were introduced to Cambridge by Forsyth's contemporary, Ernest William Hobson (1856–1933), who succeeded him in the Sadleirian Chair. Hobson had already served as President of the Philosophical Society (1906–1908) and was to be Treasurer from 1910 to 1921. His real quality as a mathematician hardly appeared till he was past the age of forty, partly no doubt because he supported himself by coaching; Miss Fawcett and John Maynard Keynes were among his pupils. (Two of the three papers he contributed to *Proceedings*—before 1900—deal with applied topics.) His great influence was exercised particularly through the successive editions of his book, *The Theory of Functions of a Real Variable* that appeared between 1907 and 1927. Exactly into the same period falls the chief work of William Henry Young (1863–1942), who contributed papers in the theory of integration (1908–11). Young, who had been a Fellow of Peter-house, though not a constant resident in the University, taught and examined over a long period. Of Hobson and Young, Hardy remarked that they were the first mathematicians in Cambridge to comprehend the importance of the French ideas of measure and integration, and he adds that it was due to their work that English mathematics lost its insularity.

Another who contributed to the modernisation of Cambridge mathematics about the turn of the century was H. F. Baker (1866–1956), whose last paper was printed when he was 85 years of age! (Baker, who long outlived Hardy, was actually ahead of Hobson in the Philosophical Society, for he was Secretary from 1896 to 1900, and President from 1902 to 1904). His *Abel's Theorem and the Allied Theory* (1897) brought to a subject long studied in Britain—the theory of algebraic func-tions—a Teutonic thoroughness acquired by its author in Göttingen, and modern continental methods. Again, E. T. Whit-taker's *Modern Analysis* (1902), a book still better known in the form perfected over a number of years by its co-author, G. N. Watson, was for many years the bible of those who wished to employ modern analysis in applied mathematics.

The geometrical work of Arthur Cayley and George Salmon was continued in Cambridge by H. W. Richmond and J. H. Grace. In 1914 H. F. Baker was appointed Lowndean Professor. A pure mathematician with interests mainly in the theory of algebraic functions, he had nevertheless followed carefully the theory of surfaces as developed by the Italian school (Castel-nuovo, Enriques, Severi) and his Presidential Address to the London Mathematical Society in 1912 had been devoted to a comprehensive survey of this topic. After his appointment Baker not only salved his conscience by delivering an annual course of lectures on dynamical astronomy, but made geometry the main subject of his teaching and research. He built up a flourishing school in this field. To begin with, his work was chiefly devoted to synthetic projective geometry and was sum-med up in the first four volumes of his *Principles of Geometry* (1922–25). Later his lectures dealt more with the theory of surfaces, concerning which he added two further volumes to his *Principles* (thus flatly contradicting the Preface to the first volume).

Among other mathematicians having a close association with the Philosophical Society early in this century were Harry

Bateman (1882–1946), author of several papers in *Transactions* between 1903 and 1910, who, after a brief association with Rutherford at Manchester (see *Proceedings*, 15, 1910) had a long and fruitful career at the California Institute of Technology from 1917 onwards; and E. W. Barnes (1874–1953), second Wrangler in his day and for over a decade a stalwart in the teaching of mathematics at Trinity, but more widely known as Bishop of Birmingham after 1924. Barnes was a Secretary from 1905 to 1912, in which office he was succeeded by G. H. Hardy (1877–1947).

Hardy has been praised by many good judges as the outstanding English mathematician of his time and universally recognized by all who knew him as one of the most fascinating of human beings. His devotion to cricket, his attention to the properties of numbers seen by chance (as on motor-cars), his eccentric ignorance of some ordinary matters of common knowledge, his passionate atheism, are remembered to this day. His autobiographical tract, *A Mathematician's Apology*, gives a fascinating picture of a social and intellectual world that is no more. His essentially Cambridge career was broken by eleven happy years at Oxford (1920–31) as Savilian Professor of Geometry; though Hardy was no original geometer himself, his important work lying in many branches of algebraic analysis, he lectured conscientiously on that part of mathematics. In both the earlier and the latter phases of his life at Cambridge Hardy contributed frequently to the Society's journals, which reflect not only his own independent interests but his collaborations with J. E. Littlewood (from 1912 onwards) and S. Ramanujan. Such work in collaboration seems to have suited Hardy particularly well. His books, particularly *A Course of Pure Mathematics* (1908) and (with E. M. Wright) *The Theory of Numbers* (1938) have had a decisive effect on the teaching of mathematics in Great Britain.

That Hardy detested the notion of the usefulness of serious mathematics is as well known as anything about him: 'No discovery of mine' he wrote, 'has made, or is likely to make, directly or indirectly, for good or ill, the least difference to the amenity of the world.' Real mathematics could of course be applied, and Hardy reckoned Maxwell, Einstein, Eddington and Dirac among real mathematicians; however, he could then regard relativity and quantum theory as cheerfully useless subjects, though he foresaw that they might not always remain so. Perhaps the last words of the *Mathematician's Apology* may, therefore, also close this brief narrative because, as it seems to me, though the Philosophical Society has at times recognized merit in utility, its main endeavour through a century and a half has been to stimulate creative thinking and research in natural philosophy, aimed not at use but knowledge. Very many of its Fellows could well have applied to themselves the austere and exacting text chosen by Hardy for himself, *mutatis mutandis*:

The case for my life, then, or for that of any one else who has been a mathematician in the same sense in which I have been one, is this: that I have added something to knowledge, and helped others to add more; and that these somethings have a value which differs in degree only, and not in kind, from that of the creations of the great mathematicians, or of any of the other artists, great or small, who have left some kind of memorial behind them.

Envoi

IT IS impossible to come to the end of a history such as this without reflecting that the functions of a scientific society and the requirements of its membership have changed greatly in a century and a half. Now, as in 1819, meetings are held which members have the privilege of attending, and publications are issued which members receive. But the realities beneath this constancy are far apart. In 1819 there were few scientific periodicals, English or foreign, and none were connected with Cambridge. Today the world is overrun with periodicals, and many Cambridge scientists are connected with their production. In 1819 there were very few men in Cambridge sincerely interested in science, and no regular way in which two or three could gather together. Today there are many hundreds of research scientists, who assemble regularly in the colloquia and tea-rooms of their laboratories, where they also enjoy specialist libraries taking in a score or many more periodicals from all over the globe. If they are not very self-denying the more senior of our Cambridge scientists attend at least one international meeting with their colleagues every year. There were no international congresses, colloquia or institutes in 1819, twelve years before the foundation of the British Association for the Advancement of Science.

The object of the Philosophical Society is, in part, the advancement of science, as it is also the advancement and stimulation of the individual Cambridge scientist; the two obviously go together. Even today a society must work largely with the same means as in 1819; the change is in the magnitude of the possibilities open to a thriving and creatively-run scientific society in an age which assigns so much more wealth to science, and which commands unparalleled techniques of communication by travel, print or electronics. Though the scene is so much more overcrowded, so to speak, in 1969 than it was in 1819, there is also a wider scope for a Cambridge Philosophical Society now than then, and it has far more resources for attaining its objectives than at any previous moment in its history.

In some respects the relationship to the Society of one of its Fellows is nowadays similar still to what it was one hundred and fifty years ago, in others it is very different. Now as then the majority of active Fellows—and all those whom the Society benefits most—are young people. At the time of the Society's foundation the problem was to interest people in science, to induce in them the desire to do scientific work themselves, and to equip them with the necessary knowledge for doing so. If science was to develop at an increasing rate it was necessary to bring more people into the scientific movement, and to reinforce the keenness of those already participating. These had been, in part, the objectives of the Manchester Literary and Philosophical Society (founded in 1781) and of the Royal Institution (1798); they were particularly the objectives of the British Association for the Advancement of Science (1831). The Yorkshire Philosophical Society, founded just two years after that at Cambridge, hoped 'to promote science in the district by establishing a scientific library, [and] scientific lectures, and by providing apparatus for original research'. From all of these the Cambridge Society differed in being a Society within a University, though not of it—always, that is to say, formally quite independent; hence it could count on great strength of local talent (as we have seen in Chapter One) but also had to face a special problem, that of preventing young abilities growing stale and idle. Only at the University of Cambridge, in the England of 1820, were a number of young men prepared, however inadequately, for a potential career as scientific investigators, and accordingly, as the *Edinburgh Review* put it in 1822:

> In Cambridge there must always be a great number of men devoted to scientific pursuits, but from want both of the facilities and the excitements furnished by such an association, apt to lose the spirit of original investigation—a remark peculiarly applicable to those young men who yearly distinguish themselves in the favourite studies of the University, and who ... are prone, if left alone, to become the mere instruments for enabling others to pursue the same course.

The possibilities and the responsibilities here seen and plainly stated had, in fact, been clearly visualized by the founders of the

Cambridge Philosophical Society and assumed by them.

A comparable situation does not exist today. If any Cambridge graduate fails now to explore the vein of originality within him for as long as it may last—ten years, twenty, more if he is lucky—neither the University nor the Philosophical Society need feel ashamed. Yet in actual fact the Society does far more at the present time than it was ever able to do in the past to promote the development of original talent. Its Library is more useful than ever before to the young research worker, who (if a Fellow or Associate) enjoys borrowing privileges as in the past. The Library has always been one of the great attractions of the Fellowship; currently nearly 500 Fellows and Associates are regular users of the Library. There are on average 60 to 70 readers each day, and nearly 3,500 loans of periodicals and parts are made during the year. Further, the Society has in recent years been in a position to aid research by making travel grants and providing translation services. It may well be able also to do more in aid of scientific publication where this presents unusual problems, as with the Arcturus volume. Under this head *Biological Reviews* has, probably, been the most important of the Society's publishing activities for more than a generation because of the high value and authority attached to its articles.

There is a sense in which a society is the journals it publishes —these are all that a member unable to attend meetings enjoys of it. From a practical point of view high quality in its journals has always been essential to the continuance of the Philosophical Society's periodical library, as was pointed out before; without exchanges the Library could not have been maintained through the years. However, it is hardly less important that journals are not only essential to the professional development of the young researcher and vital to his own self-expression, but help to make him a member of the scientific community. A society and a journal provide that mutual intellectual relationship which scientists, like other people, invariably need and for whose sake the Philosophical Society was brought into existence. From its inception its main object was the furtherance of 'scientific communication'.

Further, a great merit of such a body as the Cambridge Philosophical Society is that it can not only bring similar kinds of scientist together, but diverse kinds also. It can cross boundaries. It can cause young and old to listen to each other, experimenters and theorists, physicists and biologists. Perhaps today it is idle to dream of a truly universal philosophical society such as the men of 1819 imagined and up to a point realised. In 1819 the growing-points of science were largely descriptive—and moreover even geologists and botanists had taken the Mathematical Tripos. These growing-points, moreover, offered much of interest to the philosopher, the historian, even the theologian (and of course at that time many men, like Sedgwick, were both scientists and theologians). The early nineteenth century aimed at a synthesis of knowledge—not necessarily Whewell's omniscience—whereas our own age can at best hope to retard its fragmentation, even though it often seems that vitality is inseparable from specialisation. It would seem that one useful role for a society within a University, embracing so great and so varied abilities, might be to exploit the fact that a University brings together a large number of different scientific capacities and interests, not, as congresses do, similar ones. In such a society there is the opportunity for a true symposium. Where else can the border-line between subjects be better explored, or those techniques developed that demand mastery of more than one science? Where else should it be possible to talk about 'physics', say, or 'biology', or perhaps even about 'science', rather than minute problems?

Of course this is already happening. Many of the most active areas of modern science are interdisciplinary, such as geophysics (with its use of isotopic and geomagnetic methods), radio-astronomy, molecular biology, and the study of the properties of materials. Most scientists expect this development of new subjects on the boundaries of old ones to continue in a fruitful manner.

Possibly the Philosophical Society may be able to hasten such developments, as indeed its programmes of lectures may already be said to do. A century and a half after its foundation many new possibilities for contributing to the advancement of science lie before the Philosophical Society, and these are indeed under consideration by the Society's Council. This history has shown something of the Society's functioning and preoccupations in the past and only the future can tell what course they may now take in new conditions. For in one major respect the Philosophical Society today is unlike that of the past in that it possesses resources for the implementation of quite ambitious plans. The Society has shared in the universal prosperity that has fallen on scientific research in the last generation. One can be sure that its future plans will envisage the further development of scientific communications, and especially will not overlook the needs and potentialities of the young Cambridge men and women who are, so to speak, apprentices in scientific research. The Philosophical Society has flourished by serving the young research worker in the past and will surely continue to do so in the future.

Officers of
the Society

PRESIDENTS

1819 Rev. Will. Farish, M.A. Magd., *Jacksonian Professor*.
1821 Rev. Ja. Wood, D. D., *Master of S. John's College*.
1823 John Haviland, M. D. Joh., *Regius Professor of Physic*.
1825 Rev. Ja. Cumming, M.A., Trin., *Professor of Chemistry*.
1827 Rev. Joh. Kaye, D.D., *Master of Christ's Coll. and Bp. of Lincoln*.
1829 Rev. Tho. Turton, D.D. Cath., *Regius Professor of Divinity*.
1831 Rev. Adam Sedgwick, M.A. Trin., *Woodwardian Professor*.
1833 Rev. Joshua King, M.A. Qu., *President of Queens' College*.
1835 Rev. Will. Clark, M.D. Trin., *Professor of Anatomy*.
1837 Rev. Joh. Graham, D.D. Chr., *Master of Christ's College*.
1839 Rev. Will. Hodgson D.D. Pet., *Master of Peterhouse*.
1841 Rev. Geo. Peacock, D.D. Trin., *Lowndean Professor*.
1843 Rev. Will. Whewell, D.D. Trin., *Master of Trinity College*.
1845 Rev. Ja. Challis, M.A. Trin., *Plumian Professor*.
1847 Rev. Hen. Philpott, D.D. Cath., *Master of St. Catharine's Coll*.
1849 Rev. Rob. Willis, M.A. Gonv. and Cai., *Jacksonian Professor*.
1851 Will. Hopkins, M.A. Pet.
1853 Rev. Adam Sedgwick, M.A. Trin., *Woodwardian Professor*.
1855 Geo. Edw. Paget, M.D. Gonv. and Cai.
1857 Will. Hallows Miller, M.D. Joh., *Professor of Mineralogy*.
1859 Geo. Gabriel Stokes, M.A. Pemb., *Lucasian Professor*.
1861 Joh. Couch Adams, M.A. Pemb., *Lowndean Professor*.
1863 Will. Hepworth Thompson, M.A. Trin., *Regius Professor of Greek*.
1865 Rev. Hen. Wilkinson Cookson, D.D. Pet., *Master of Peterhouse*.
1867 Rev. Will. Selwyn, D.D. Joh., *Lady Margaret's Professor*.
1869 Art. Cayley, M.A. Trin., *Sadleirian Professor*.
1871 Geo. Murray Humphry, M.D. Down., *Professor of Anatomy*.
1873 Ch. Cardale Babington, M.A. Joh., *Professor of Botany*.
1875 Ja. Clerk Maxwell, M.A. Trin., *Cavendish Professor of Experimental Physics*.

1877   Geo. Downing Liveing, M.A. Joh., *Professor of Chemistry.*

1879   Alf. Newton, M.A. Magd., *Professor of Zoology and Comparative Anatomy.*

1881   Fra. Maitland Balfour, M.A. Trin.

1882   Ja. Whitbread Lee Glaisher, M.A. Trin.

1884   Mich. Foster, M.A. Trin., *Professor of Physiology.*

1886   Rev. Coutts Trotter, M.A. Trin.

1888   Joh. Willis Clark, M.A. Trin.

1890   G. H. Darwin, *Plumian Professor, later Kt.*

1892   T. McKenny Hughes, *Professor of Geology.*

1894   J. J. Thomson, *Cavendish Professor, later Kt.*

1896   Francis Darwin, *Reader in Botany, later Kt.*

1898   Joseph Larmor, *later Lucasian Professor and Kt.*

1900   A. Macalister, *Professor of Anatomy.*

1902   H. F. Baker, *later Lowndean Professor.*

1904   H. Marshall Ward, *Professor of Botany.*

1906   E. W. Hobson, *later Sadleirian Professor.*

1908   Adam Sedgwick, *Professor of Zoology,* 1907–1909.

1909   William Bateson, *Professor of Biology*

1910   Sir George Darwin, *Plumian Professor,*

1912   A. E. Shipley, *Master of Christ's and Reader in Zoology, later Kt.*

1914   H. F. Newall, *Director, Solar Physics Observatory.*

1916   J. E. Marr, *later Professor of Geology.*

1918   C. T. R. Wilson, *later Jacksonian Professor.*

1920   A. C. Seward, *Master of Downing and Professor of Botany, later Kt.*

1922   C. T. Heycock, *Reader in Metallurgy.*

1924   J. T. Wilson, *Professor of Anatomy.*

1926   Horace Lamb, *Rayleigh lecturer, later Kt.*

1928   G. Udny Yule, *Lecturer in Statistics.*

1930   F. M. J. Stratton, *Professor of Astrophysics.*

1931   A. Hutchinson, *Master of Pembroke and Professor of Mineralogy.*

1933   Joseph Barcroft, *Professor of Physiology, later Kt.*

1935   F. W. Aston, *Trinity College, Mass-spectrographer.*

1937   Sir F. G. Hopkins, *Professor of Biochemistry.*

1939   W. H. Mills, *Reader in Stereochemistry.*

1941   James Gray, *Professor of Zoology, later Kt.*

1943   E. K. Rideal, *Professor of Colloid Science, later Kt.*

1945   F. T. Brooks, *Professor of Botany.*

1947   W. V. D. Hodge, *Lowndean Professor, later Kt.*

1949   W. B. R. King, *Professor of Geology.*

1951   H. J. Emeléus, *Professor of Inorganic Chemistry.*

1953   E. C. Bate-Smith, *Director, Low Temperature Research Station.*

1955   F. J. W. Roughton, *Professor of Colloid Science.*

1957   F. P. Bowden, *later Professor of Surface Physics.*

1959   E. J. H. Corner, *later Professor of Tropical Botany.*

1961   F. P. White, *St. John's College, Lecturer in Mathematics.*

1963 E. N. Willmer, *later Professor of Histology.*

1965 Sir Rudolph Peters, *formerly Professor of Biochemistry, University of Oxford.*

1967 Sir Geoffrey Taylor, *formerly Royal Society Research Professor.*

1968 J. A. Ratcliffe, *formerly Reader in Ionospheric Physics.*

SECRETARIES.

1819 Rev. Adam Sedgwick, M.A. Trin. *Professor of Geology.*
Rev. Sam. Lee, M.A. Qu., *Professor of Arabic.*

1821 Rev Geo. Peacock, M.A. Trin.
Joh. Stevens Henslow, M.A. Joh.

1826 Rev. Joh. Stevens Henslow, M.A. Joh., *Professor of Mineralogy and Botany.*
Rev. Will. Whewell, M.A. Trin.

1833 Rev. Joh. Stevens Henslow, M.A. Joh., *Professor of Botany.*
Rev. Will. Whewell, M.A. Trin.
Rev. Joh. Lodge, M.A. Magd., *University Librarian.*

1836 Rev. Joh. Stevens Henslow, M.A. Joh., *Professor of Botany.*
Rev. Will. Whewell, M.A.Trin.
Rev. R. Willis, M.A. Gonv. and Cai.

1839 Rev. Will. Whewell, M.A. Trin.
Rev. R. Willis, M.A. Gonv. and Cai., *Jacksonian Professor.*
Will. Hopkins, M.A. Pet.

1842 Rev. R. Willis, M.A. Gonv. and Cai., *Jacksonian Professor.*
Will. Hopkins, M.A. Pet.
Will. Hallows Miller, M.D. Joh., *Professor of Mineralogy.*

1851 Will. Hallows Miller, M.D. Joh., *Professor of Mineralogy.*
Ch. Cardale Babington, M.A. Joh.
Geo. Gabriel Stokes, M.A. Pemb., *Lucasian Professor.*

1854 Ch. Cardale Babington, M.A. Joh.
Joh. Couch Adams, M.A. Pemb.
Rev. Ch. Fre. Mackenzie, M.A. Gonv. and Cai.

1855 Ch. Cardale Babington, M.A. Joh.
Joh. Couch Adams. M.A. Pemb.
Geo. Downing Liveing, M.A. Joh.

1858 Ch. Cardale Babington, M.A. Joh.
Geo. Downing Liveing, M.A. Joh.
Norman Macleod Ferrers, M.A. Gonv. and Cai.

1866 Ch. Cardale Babington, M.A. Joh., *Professor of Botany.*
Geo. Downing Liveing, M.A. Joh., *Professor of Chemistry.*
Rev. T. G. Bonney, M.A. Joh.

1870 Rev. T. G. Bonney, M.A. Joh.
Joh. Willis Clark, M.A. Trin.
Rev. Coutts Trotter, M.A. Trin.

1873 Joh. Willis Clark, M.A. Trin.

1873   Rev. Coutts Trotter, M.A. Trin.
       Rev. Joh. Batteridge Pearson, B.D. Emm.
1878   Joh. Willis Clark, M.A. Trin.
       Rev. Coutts Trotter, M.A. Trin.
       Ja. Whitbread Lee Glaisher, M.A. Trin.
1882   Joh. Willis Clark, M.A. Trin.
       Rev. Coutts Trotter, M.A. Trin.
       Will Mitchinson Hicks. M.A. Joh.
1883   Rev. Coutts Trotter, M.A. Trin.
       Ri. Tetley Glazebrook, M.A. Trin.
       Sydney Howard Vines, M.A. Chr.
1886   Ri. Tetley Glazebrook, M.A. Trin.
       Sydney Howard Vines, M.A. Chr.
       Jos Larmor, M.A. Joh.
1887   Sydney Howard Vines, M.A. Chr.
       Jos. Larmor, M.A. Joh.
       Matth. Moncrieff Pattison-Muir, M.A. Gonv. and Cai.
1888   Jos. Larmor, M.A. Joh.
       Matth. Moncrieff Pattison-Muir, M.A. Gonv. and Cai.
       Sidney Fre. Harmer, M.A. King's.
1889   Jos. Larmor, M.A. Joh.
       Sidney Fre. Harmer, M.A. King's.
       Andr. Russell Forsyth, M.A. Trin.
1890   Joseph Larmor, Joh. (*President, 1898*)
       S. F. Harmer, King's.
       E. W. Hobson, Chr. (*President, 1906*)
1891   No change.
1892   E. W. Hobson, Chr.
       Joseph Larmor, Joh.
       W. Bateson, Joh. (*President, 1909*)
1893   Joseph Larmor, Joh.
       H. F. Newall, Trin. (*President, 1914*)
       W. Bateson, Joh.
1894   No change.
1895   No change.
1896   H. F. Newall, Trin.
       W. Bateson, Joh.
       H. F. Baker, Joh. (*President, 1902*)
1897   No change.
1898   No change.
1899   H. F. Baker, Joh., (*President, 1902*)
       A. E. Shipley, Chr. (*President, 1912*)
       L. R. Wilberforce, Trin.
1900   H. F. Baker, Joh.
       A. E. Shipley, Chr.
       S. Skinner, Chr.

1901	A. E. Shipley, Chr.
	S. Skinner, Chr.
	H. M. Macdonald, Clare.
1902	No change.
1903	No change.
1904	A. E. Shipley, Chr.
	G. B. Mathews, Joh.
	P. V. Bevan, Trin.
1905	A. E. Shipley, Chr.
	E. W. Barnes, Trin.
	P. V. Bevan, Trin.
1906	No change.
1907	No change.
1908	A. E. Shipley, Chr.
	E. W. Barnes, Trin.
	Alex. Wood, Emm.
1909	No change.
1910	No change.
1911	E. W. Barnes, Trin.
	Alex. Wood, Emm.
	F. A. Potts, Trin. H.
1912	Alex. Wood, Emm.
	F. A. Potts, Trin. H.
	G. H. Hardy, Trin. (*later Sadleirian Professor*)
1913	No change.
1914	No change
1915	Alex. Wood, Emm.
	E. A. N. Arber, Trin.
	G. H. Hardy, Trin.
1916	Alex. Wood, Emm.
	G. H. Hardy, Trin.
	H. H. Brindley, Joh.
1917	No change.
1918	No change.
1919	No change.
1920	H. H. Brindley, Joh.
	H. F. Baker, Joh. (*President, 1902*)
	F. W. Aston, Trin. (*President, 1935*)
1921	No change.
1922	H. F. Baker, Joh.
	F. W. Aston, Trin.
	James Gray, King's (*President, 1941*)
1923	No change.
1924	F. W. Aston, Trin.
	James Gray, King's
	F. P. White, Joh. (*President, 1961*)

1925 James Gray, King's (*President, 1941*)
      F. P. White, Joh.
      R. H. Fowler, Trin. (*later Plummer Professor and Kt.*)

1926 F. P. White, Joh.
      R. H. Fowler, Trin.
      H. Munro Fox, Gonv. and Cai. (editor of *Biological Reviews,* 1926–1967)

1927 F. P. White, Joh.
      R. H. Fowler, Trin.
      F. T. Brooks, Emm. (*President, 1945*)

1928 F. P. White, Joh.
      F. T. Brooks, Emm.
      D. R. Hartree, Joh. (*later Plummer Professor*)

1929 F. P. White, Joh.
      W. B. R. King, Jesus, (*President, 1949*)
      J. D. Cockcroft, Joh. (*later Jacksonian Professor, Master of Churchill and Kt.*)

1930 No change.

1931 F. P. White, Joh.
      J. D. Cockcroft, Joh.
      H. Hamshaw Thomas, Down. (*later Reader in Botany*)

1932 No change.
1933 No change.
1934 No change.
1935 No change.

1936 J. D. Cockcroft, Joh.
      A. H. Wilson, Trin. (*later Kt.*)
      O. M. B. Bulman, Sid. (*later Professor of Geology*)

1937 A. H. Wilson, Trin.
      O. M. B. Bulman, Sid.
      J. A. Ratcliffe, Sid. (*later Reader in Physics*)

1938 No change.

1939 A. H. Wilson, Trin.
      O. M. B. Bulman, Sid.
      R. C. Evans, Cath.

1940 No change.

1941 A. H. Wilson, Trin.
      O. M. B. Bulman, Sid.
      W. V. D. Hodge, Pemb. (*President, 1947*)

1942 No change.
1943 No change.
1944 No change.

1945 M. S. Bartlett, Queens'.
      E. S. Shire, King's.
      J. E. Smith, Jesus.

1946 M. S. Bartlett, Queens'.

1946 R. C. Evans, Cath.
      J. E. Smith, Jesus.
1947 R. C. Evans, Cath.
      J. E. Smith, Jesus.
      R. A. Rankin, Clare.
1948 No change.
1949 No change.
1950 R. A. Rankin, Clare.
      Sydney Smith, Cath.
      A. B. Pippard, Clare (*later Plummer Professor*)
1951 Sydney Smith, Cath.
      A. B. Pippard, Clare.
      D. R. Taunt, Jesus.
1952 No change.
1953 No change.
1954 Sydney Smith, Cath.
      A. B. Pippard, Clare.
      H. G. Eggleston, Trin.
1955 Sydney Smith, Cath.
      H. G. Eggleston, Trin.
      T. E. Faber, Trin.
1956 No change.
1957 Sydney Smith, Cath.
      T. E. Faber, Trin.
      H. P. F. Swinnerton-Dyer, Trin.
1958 No change.
1959 No change.
1960 Sydney Smith, Cath.
      T. E. Faber, Trin.
      J. H. Williamson, Chr.
1961 No change.
1962 Sydney Smith, Cath.
      J. H. Williamson, King's.
      J. S. Griffith, King's
1963 Sydney Smith, Cath.
      T. G. Murphy, Joh.
      A. D. Yoffe, Darwin.
1964 No change
1965 Sydney Smith, Cath.
      G. R. Allan, Chur.
      .L. T. Chadderton, Caius
1966 Sydney Smith, Cath.
      G. R. Allan, Chur.
      J. E. Field, Down.
1967 J. E. Field, Down.
      C. R. F. Maunder, Chr.

1967   M.A. Message, Cath.
1968   No change.

TREASURERS.

1819   Rev. Bewick Bridge, B.D. Pet.
1825   Fre. Thackeray, M.D. Emm.
1834   Rev. Geo. Peacock, M.A. Trin.
1839   Geo Edw. Paget, M.D. Gonv. and Cai.
1853   Rev. Tho. Hedley, M.A. Trin.
1857   Rev. Will. Magan Campion, M.A. Qu.
1876   Ja. Whitbread Lee Glaisher, M.A. Trin.
1878   Rev. Joh. Batteridge Pearson, D.D. Emm.
1883   Joh. Willis Clark, M.A. Trin.
1887   R. T. Glazebrook, Trin.
1898   A. E. Shipley, Chr.
1899   H. F. Newall, Trin.
1910   E. W. Hobson, Chr.
1921   F. A. Potts, Trin. H.
1937   J. D. Cockcroft, Joh.
1946   M.V. Wilkes, Joh.
1958   R. C. Evans, Cath.

Additional Biographical Notes on Some Officers of the Society

I. PRESIDENTS

ASTON, Francis William (1877–1945) brought much experience with him when he came to Cambridge in 1910 to be J. J. Thomson's research assistant. His first mass-spectrograph (developed from the positive-ray apparatus) was built in 1919; with this he began the complete investigation of isotopy. He was elected F.R.S. in 1921, and was the Society's President from 1935 to 1937.

BAKER, Henry Frederick (1866–1956) was President from 1902 to 1904 and contributed many papers to the Society. He had previously served as Secretary (1896–1900) and was Lowndean Professor from 1914 to 1936.

BARCROFT, Sir Joseph (1872—1947) succeeded John Newport Langley as Professor of Physiology in 1925. His work on haemoglobin, on the respiratory function of the blood, on the spleen, and on embryonic physiology was all of high importance. He was President from 1933 to 1935.

BOWDEN, Frank Philip (1903–68), President 1957–59, established solid-state physics as an independent subject in Cambridge and also gave much service to industry.

BROOKS, Frederick Tom (1882–1952), Professor of Botany from 1937 to 1948, was Secretary in 1927 and President from 1945 to 1947. His interest in mycology and plant pathology, on which he contributed only three out of the many papers he published to the Society, was stimulated by Marshall Ward.

GLAISHER, James Whitbread Lee (1848–1928), President from 1882 to 1884, was a Fellow of Trinity from 1871 to 1928 and a College Lecturer from 1871 to 1901. Author of nearly 400 papers on pure mathematics and their applications to astronomy, he was active in both the London Mathematical and the Royal Astronomical Societies. He had also served as Secretary and Treasurer of the Philosophical Society.

HEYCOCK, Charles Thomas (1858–1931), President from 1922 to 1924, was a Fellow of King's and Reader in Metallurgy from 1908 to 1929.

HOPKINS, Sir Frederick Gowland (1861–1947) was brought from London by Michael Foster to teach chemical physiology in 1898. He was also a tutor at Emmanuel. He was one of the first to demonstrate the critical dietetic role of vitamins (1912). For this work he shared the Nobel Prize in 1929. He became Professor of biochemistry in 1914 and was President of the Society 1937–39.

HUGHES, Thomas McKenny (1831–1917), Woodwardian Professor from 1873 to 1917, was one of the great teachers of geology. He was President 1892–94 and author of many contributions on geology.

HUTCHINSON, Arthur (1866–1937) taught mineralogy in Cambridge 1895–1931 (Professor, 1926). He was President, 1931–33.

KING, William Bernard Robinson (1889–1963), after distinguished war service rejoined Professor J. E. Marr in the Geology Department at Cambridge, where he held the Woodwardian Chair from 1943 to 1955. From 1931 to 1943 he had occupied the Chair at University College, London. He was Secretary 1929–30 and President 1949–51.

LAMB, Sir Horace (1849–1934), Fellow of Trinity, great authority on mechanics, contributed one paper in 1900 and others after his return to Cambridge from his chair at Manchester in 1920. He was President from 1926 to 1928.

MACALISTER, Alexander (1844–1919) graduated at Trinity College, Dublin, where he was Professor of Zoology from 1869 to 1877, and of Anatomy and Surgery, 1877–83. In 1883 he moved to the Cambridge Chair of Anatomy, which he held for the rest of his life. He was President 1900–1902.

MARR, John Edward (1857–1933) took the Natural Sciences Tripos in 1878 and became a Fellow of St. John's in 1881. From 1886 to 1917, when he succeeded him in the Woodwardian Chair, Marr taught under T. McK. Hughes. As an authority in palaeozoic rocks and a great teacher, Marr helped raise the Cambridge school of geology to eminence. He was President 1916–18.

MILLS, William Hobson (1873–1959), after starting research in the Chemistry Department in 1897, and teaching at the Northern Polytechnic in London (1902–12) returned to Cambridge first as Demonstrator, then as Lecturer (1919) and finally as Reader in Stereochemistry (1931–38). He did much work on cyanine dyes, which proved of great importance in making panchromatic plates for aerial photography in the 1914–18 War. His stereochemical work also was of great scientific value. He was President from 1939 to 1941.

NEWALL, Hugh Frank (1857–1944) returned to Cambridge in 1885 to assist J. J. Thomson at the Cavendish. After his father's death in 1891 he brought the 25-inch 'Newall' Telescope to Cambridge, with which he worked for the next 50 years. He held (without stipend) the new Chair of Astrophysics 1901–1928, and was Director

Sir A. C. Seward 1863–1941

of the Solar Physics Observatory from 1913. He was Secretary 1893–1898, Treasurer of the Society from 1899 to 1910, and President 1914–16.

SEWARD, Sir Albert Charles (1863–1941) took up palaeobotany in 1886; he taught botany in Cambridge from 1890 to 1936 (Professor, 1906), and was Master of Downing (1915–36). He made the Botany School supremely successful. He was President 1920–22 and contributed numerous papers, 1890–1906.

STRATTON, Frederick John Marrian (1881–1960), whose interest in astrophysics began about 1910, succeeded Newall as Professor and Director of the Solar Physics Laboratory (1928–47). He was elected President in 1930, but was incapacititated by temporary ill-health in the following year.

TROTTER, Coutts (1837–87), Secretary 1870–83, President 1886, lectured on physical science at Trinity from 1869 to 1884. He had previously studied with Helmholtz and Kirchoff. He became a very influential figure in Cambridge, and a benefactor to his College.

WARD, Harry Marshall (1854–1906) studied under T. H. Huxley and at Manchester before coming to Cambridge, and later worked with Sachs. After many investigations and appointments he returned to Cambridge as successor to Babington in the Chair of Botany (1895). He rebuilt the Botany School both literally and intellectually. Ward was President 1904–06.

WILSON, James Thomas (1861–1945), President from 1924 to 1926, was Professor of Anatomy from 1920 to 1934; previously he had graduated and taught in Edinburgh and occupied a Chair at Sydney, Australia.

II. OTHER OFFICERS

BARNES, Ernest William (1874–1953), Second Wrangler and first Smith's prizeman, taught mathematics at Trinity until 1915. He was Bishop of Birmingham from 1924 onwards. He was Secretary from 1905 to 1911 inclusive and a contributor of major papers to *Transactions*.

BEVAN, Penry Vaughan (1875–1913), Fellow of Trinity and University Demonstrator under J. J. Thomson, was Professor of Physical Science at Royal Holloway College, 1908–13. He was Secretary 1904–1907 inclusive and a frequent contributor on experimental physics.

BRINDLEY, Harold Hulme (1865–1944), Fellow of St. John's, taught biology in Cambridge from 1915 to 1934. He was Secretary 1916–21 inclusive, and gave the Society a number of papers, chiefly on the common earwig.

COCKCROFT, Sir John Douglas (1897–1967) worked under Chadwick at the Cavendish from 1924. Next he helped Kapitza produce high magnetic fields. At the end of 1928 he began (with E. T. S. Walton) the building of a 300 KV proton accelerator; later a 500-600 KV machine was built with which (in 1932) alpha particles were produced from a lithium target. Later still he designed and built the 37 inch cyclotron. He was brought into radar shortly before the war, and this provided his main work until 1944, when he was involved in research on the bomb. In 1946 he was appointed Director of the Atomic Energy Research Establishment at Harwell, whence he returned to Cambridge as Master of Churchill College in 1959. Cockcroft was a Secretary from 1929 to 1937 and Treasurer from 1937 to 1946.

FOWLER, Sir Ralph Howard (1889–1944) did much official work in both wars; his main research was in statistical mechanics and atomic physics, on which he offered many papers to the Society. He was first Plummer Professor of Mathematical Physics (from 1932). He was Secretary 1925–27.

FOX, Harold Munro (1889–1967) studied at Cambridge before the 1914 War and taught there after it. He was Professor of Zoology at Birmingham University from 1927 to 1941, and at Bedford College, London, 1941–54. He edited *Biological Reviews* from 1926 to 1967.

GLAZEBROOK, Sir Richard Tetley (1854–1935) began research under Maxwell, acted as demonstrator for Rayleigh, became assistant director of the Cavendish from 1891. He was the first Director of the National Physical Laboratory (1899), where he had great personal effect on aeronautical research (from 1909). While at Cambridge he contributed many papers to the Society and acted as Treasurer, 1887–98.

HARMER, Sir Sidney Frederic (1862–1950), Fellow of King's 1886–1914, taught zoology in Cambridge from 1883–1908, when he moved to the British Museum (Natural History). He was President of the Linnean Society, and Secretary of the Philosophical Society from 1888 to 1891 inclusive and contributed several papers.

HARTREE, Douglas Rayner (1897–1958), whose undergraduate career was broken by the 1914–18 War, was a Fellow of St. John's and Christ's from 1924 to 1929, in which period he published in the *Proceedings* contributions of first-rate importance to quantum physics based on his theory of the self-consistent field. From 1934 he began work with the differential analyzer, and after the war played a major role in the development of electronic computer techniques. He was Professor at Manchester from 1929, until in 1946 he returned to the Plummer Chair of Mathematical Physics.

MACDONALD, Hector Munro (1865–1935), Fellow of Clare 1890–1908, Professor of Mathematics at Aberdeen from 1904. He was Secretary 1901–1903 inclusive.

MATHEWS, George Ballard (1861–1922), Senior Wrangler in 1883, was Professor of Mathematics at Bangor, 1884–96, and afterwards lecturer at Cambridge and the University College of North Wales. He was Secretary in 1904.

MUIR, Matthew Moncrieff Pattison (1848–1931), Secretary in 1887 and 1888, was a Life-Fellow of Caius; from 1877 to 1908 he was in charge of the College's chemical laboratory.

PEARSON, John Batteridge (1832–1918), Fellow of Emmanuel 1856–83, held a Cornish living 1883–1912. He was Secretary in 1873, Treasurer 1878–83. He contributed numerous papers on astronomy, 1875–86.

POTTS, Frank Armitage (c. 1882–1937) was a member of the Zoology Department for twenty-five years. He was a Fellow of Trinity Hall from 1908 and for many years Praelector. He was Secretary from 1911 to 1915 and Treasurer of the Society from 1921 to 1937—difficult years—before which time he had contributed several papers on crustacea.

THOMAS, Hugh Hamshaw (1885–1962), Secretary of the Society from 1931 to 1935 inclusive, published a number of papers in the *Proceedings* on palaeobotany. He was Reader in Botany from 1937 to 1950.

VINES, Sydney Howard (1849–1934) Secretary from 1883 to 1887, having studied with T. H. Huxley and W. T. Thistleton-Dyer, introduced modern botany to Cambridge in his lectures at Christ's from 1876 onwards. He became a Reader in Botany in 1881, but moved to the Oxford Chair in 1888, after which he achieved little.

WILBERFORCE, Lionel Robert (1861–1944) demonstrated for J. J. Thomson between 1887 and 1900; from 1900 to 1935 he was Professor of Physics at Liverpool. He contributed papers on electricity, and was a Secretary in 1899.

WOOD, Alex (1879–1950), Secretary from 1908 to 1920, worked in the Cavendish from 1902, and was a University Lecturer in Physics from 1920 to 1944, as well as being very active in College, University and town affairs.

Index